Population Ecology
of Individuals

MONOGRAPHS IN POPULATION BIOLOGY

EDITED BY ROBERT M. MAY

1. The Theory of Island Biogeography, by Robert H. MacArthur and Edward O. Wilson
2. Evolution in Changing Environments: Some Theoretical Explorations, by Richard Levins
3. Adaptive Geometry of Trees, by Henry S. Horn
4. Theoretical Aspects of Population Genetics, by Motoo Kimura and Tomoko Ohta
5. Populations in a Seasonal Environment, by Stephen D. Fretwell
6. Stability and Complexity in Model Ecosystems, by Robert M. May
7. Competition and the Structure of Bird Communities, by Martin L. Cody
8. Sex and Evolution, by George C. Williams
9. Group Selection in Predator-Prey Communities, by Michael E. Gilpin
10. Geographic Variation, Speciation, and Clines, by John A. Endler
11. Food Webs and Niche Space, by Joel E. Cohen
12. Caste and Ecology in the Social Insects, by George F. Oster and Edward O. Wilson
13. The Dynamics of Arthropod Predator-Prey Systems, by Michael P. Hassell
14. Some Adaptations of Marsh-Nesting Blackbirds, by Gordon H. Orians
15. Evolutionary Biology of Parasites, by Peter W. Price
16. Cultural Transmission and Evolution: A Quantitative Approach, by L. L. Cavalli-Sforza and M. W. Feldman
17. Resource Competition and Community Structure, by David Tilman
18. The Theory of Sex Allocation, by Eric L. Charnov
19. Mate Choice in Plants: Tactics, Mechanisms, and Consequences, by Mary F. Wilson and Nancy Burley
20. The Florida Scrub Jay: Demography of a Cooperative-Breeding Bird, by Glen E. Woolfenden and John W. Fitzpatrick
21. Natural Selection in the Wild, by John A. Endler
22. Theoretical Studies on Sex Ratio Evolution, by Samuel Karlin and Sabin Lessard
23. A Hierarchical Concept of Ecosystems, by R. V. O'Neill, D. L. DeAngelis, J. B. Waide, and T. F. H. Allen
24. Population Ecology of the Cooperatively Breeding Acorn Woodpecker, by Walter D. Koenig and Ronald L. Mumme
25. Population Ecology of Individuals, by Adam Łomnicki

Population Ecology of Individuals

ADAM ŁOMNICKI

PRINCETON UNIVERSITY PRESS

PRINCETON, NEW JERSEY

1988

Library of Congress Cataloging in Publication Data will be found on the
last printed page of this book
ISBN 0-691-08471-8 (cloth)   0-691-08462-9 (pbk.)
This book has been composed in Linotron Baskerville
Clothbound editions of Princeton University Press books are printed on
acid-free paper, and binding materials are chosen for strength and
durability. Paperbacks, although satisfactory for personal collections, are not
usually suitable for library rebinding
Printed in the United States of America by Princeton University Press,
Princeton, New Jersey

Contents

Preface ix

1. Introduction: Basic Models of Population Ecology
   and Intrapopulation Variability 1
   1.1. Individuals and Superorganisms 1
   1.2. Unlimited Population Growth 5
   1.3. Limited Population Growth 11
   1.4. Limited Growth in Discrete Places 15

2. Individual Variation in Resource Partitioning and
   Population Dynamics 20
   2.1. Resource Partitioning among Individuals 20
   2.2. Four Versions of the Model of Resource
        Partitioning 25
   2.3. Population Stability and Persistence 30
   2.4. Laboratory and Field Data 34
   2.5. Scramble and Contest Competition 42

3. Individual Variation of Body Weight in Plant and
   Animal Populations 46
   3.1. Empirical Data 46
   3.2. Some Simple Explanations and Their
        Shortcomings 48
   3.3. Weight Differentiation under Stochastic
        Growth 53
   3.4. Deterministic Growth and the Importance of
        Early Differentiation 57
   3.5. Weight Differentiation in Competition for
        Space 59
   3.6. Weight Distribution and General Properties
        of the Function $y(x)$ 61

v

4. Individual Differences and Hereditary Variation   65

   4.1. Variation as an Adaptation   66
   4.2. Differential Mortality and the Soft Selection
      Concept   72
   4.3. Genetic Determination of Individual Success
      in the Ecological World   81

5. Age and Overlapping Generations   86

   5.1. Age-Dependent Individual Success   86
   5.2. Distinct Life Stages within a Population   91
   5.3. Simple Extension of the Models of
      Population Dynamics to Overlapping
      Generations   95
   5.4. Discrete Versus Continuous Models of
      Population Dynamics   100

6. The Mechanism of Contest Competition   106

   6.1. Definition of Contest Competition   107
   6.2. Population Effects of Contest Competition   111
   6.3. The Case Study: Competition among Gall
      Aphids *Pemphigus betae*   116
   6.4. Social Hierarchy, Territoriality, and Contest
      Competition   118
   6.5. Evolutionarily Stable Arms Investments, or
      How Contest Competition Can Be Regarded
      as a Result of Arms Races   120

7. Self-regulation of Population Size   124

   7.1. Self-regulation in Terms of Game Theory   126
   7.2. Self-regulation in Confined Laboratory
      Populations   132
   7.3. Optimal Reproduction in Populations with
      Unequal Resource Partitioning   136

8. Emigration and Unequal Resource Partitioning   143

   8.1. Emigration from Groups of Related and
      Unrelated Individuals   143

8.2. Impermanent Local Habitats in
Heterogeneous Space   150
8.3. Evolution of Emigration from Local
Populations without Unequal Resource
Partitioning   155
8.4. Emigration from Local Populations with
Unequal Resource Partitioning   161
8.5. Emigration, and Scramble and Contest
Competition   166
8.6. Free and Despotic Distribution of Animals   170

9. Field and Laboratory Populations of Animals   174
9.1. Limitations of Field Studies   174
9.2. Animal Populations in the Laboratory   176
9.3. Free and Confined Laboratory Populations
of *Hydra*   178
9.4. Free and Confined Laboratory Populations
of Flour Beetles   183
9.5. Confined Populations of Animals in the Field   186

10. Spatial and Temporal Heterogeneity and Stability
of Ecological Systems   189
10.1. Reproduction in Spatially and Temporally
Heterogeneous Environments   190
10.2. Spatial Microheterogeneity   194
10.3. Direct Relations between Spatial
Heterogeneity, Individual Variability, and
Stability   197
10.4. Spatial Heterogeneity and Species Diversity   199
10.5. Spatial Heterogeneity and Ecosystem
Stability   201

References   205

Author Index   217

Subject Index   220

Preface

This book has been written with the conviction that further progress in ecology requires taking into account the fact that ecological systems are made up of individuals that differ among themselves, and not only in their taxonomic affiliation, sex, and age. The idea that individuals differ is not a new one, but for a long time attempts to develop ecological theory have been made without taking this fact into account. Individual variation seemed to be nothing more than an unimportant hindrance in the study of the structure and stability of ecological systems. In this book I shall attempt to convince the reader that ecology, like other parts of biology, should apply a reductionist approach more consistently, by deriving the properties of ecological systems from the properties of their elements, i.e. individuals, and that for this an understanding of variation among individuals is essential. Biologists are well aware of the importance of hereditary variation, especially where the theory of evolution is concerned, but the importance of any individual variation, both hereditary and environmental, is only now becoming recognized in ecological theory.

By applying some simple mathematical models, I will try to show that within population variation in resource intake may explain certain phenomena that until now have been poorly understood and that not so long ago were still explained by the action of group selection. More detailed theoretical analyses of individual variation can lead to a better understanding of such problems as, for example, the mechanism of intraspecific competition; but on the other hand these analyses give rise to new questions. For example, I think that we have not yet fully understood the relation between contest competition and the dispersal behavior of animals in nature.

PREFACE

I began to write this book during my stay at Carleton College, Northfield, Minnesota, and I have finished it due to encouragement from Robert M. May, who suggested the possibility of publishing it in the Princeton Monographs series. Early drafts of the first three chapters were read by Paul Jensen, Bruce R. Levin, Richard E. Lenski, Edward B. Swain, Jonathan Brown, Janusz Uchmanski, and January Weiner; the complete manuscript was read by Michał Jasienski, Jan Kozłowski, Danuta Padley, and two reviewers: Michael P. Hassell and H. Ronald Pulliam. The suggestions and criticisms I have received were a great help in my attempts to improve the manuscript.

Population Ecology
of Individuals

Introduction: Basic Models of Population Ecology and Intrapopulation Variability

1.1. INDIVIDUALS AND SUPERORGANISMS

Although ecology and the theory of natural selection are parts of biology, their concepts and methods are far removed from those of other areas of biology. Traditional biology has been concerned with identifying many different kinds of plants and animals, describing their morphology and anatomy; more recently, it has been concerned with physiological and biochemical processes within a single organism. To understand the relations among individuals, which are of fundamental importance in ecology and the theory of natural selection, requires different approaches and methods.

Gould's (1980) account of how Charles Darwin formulated the theory of natural selection is convincing evidence that traditional biology had not provided its students with the methods to study population processes. It seems that Darwin's enormous biological knowledge alone was not sufficient for him to grasp clearly the mechanisms of evolution. Only after reading Thomas Malthus' "An Essay on the Principle of Population" and Dugald Stewart's "On the Life and Writing of Adam Smith," as well as some statistical articles by Adolph Quetelet, was he able to understand these mechanisms and to formulate his theory. He obviously required a knowledge of population processes that in

1

those days was available only in books on economics, demography, and statistics.

Biologists who started to practice ecology at the beginning of our century did not look for inspiration in economics and statistics; they were relatively uninterested in the theory of natural selection. Rather, they attempted to develop ecological theory by applying methods specific to the study of separate organisms. But they did not try to predict the properties of populations or communities from the properties of organisms that belong to these large units. Such an approach was represented by a rather narrow group of theoreticians: V. Volterra, A. J. Lotka, and V. A. Kostitzin. Most ecologists of those days were fascinated by the idea of ecological systems as kinds of superorganisms, with properties analogous to those of individual organisms.

The enormous diversity of relations among the many different species of plants and animals living in a forest or in a lake can discourage anyone who might be tempted to give a detailed description of such systems. On the other hand, biologists have been relatively successful in understanding a single organism, or at least in predicting its behavior. If forests and lakes are kinds of large organisms, it should be possible to describe them, to find out how they work, and to predict their behavior by applying the concepts developed for studying separate organisms. When reading old ecological textbook such as that by Allee et al. (1950), or studying the phytosociological ideas of Braun-Blanquet (1932), one can see just how common was the concept of ecological units as kinds of superorganisms. Nor was such an image limited to ecology; a similar approach was applied to the theory of natural selection, in which not only individuals but also populations, species, or even ecosystems have been regarded as the units of selection. Unfortunately, such views were rarely expressed explicitly, which made them difficult to criticize. The most explicit and important presentation of the concept of group selection, i.e. selection acting above the level of individuals, was published by Wynne-Edwards (1962). That such an idea was stated explicitly must be seen as a great con-

2

tribution to evolutionary and ecological theory. The critique that followed has made it possible to abandon the concept of group selection and, consequently, the concept of superorganism in biology. Criticism of this concept can also be seen in more recent ecological textbooks (e.g., Colinveaux [1973]).

Progress in evolutionary biology during the last decades has enabled ecologists to see that within the hierarchy from cell to tissue, to individual organism, to population, to community, and finally to biosphere, the individual organism is something distinct. Within this hierarchy, the systems ranging from sexually reproducing individuals down to cells, are sets of genetically identical elements, whereas those ranging from groups up to ecosystems are sets of genetically different individuals. Even if one assumes after Dawkins (1982) that gene-replicators, not individuals, are the units of selection, we can still regard an individual as an entity that is adapted to survive and to reproduce. Neither populations of sexually reproducing individuals nor individual cells are adapted by natural selection, as individuals are. Where such adaptation can be found, it is an exception that occurs under special circumstances, as in the case of clones (sections 7.1 and 9.3).

The genetic uniqueness of individuals has important ecological consequences. One can say that cells and tissues within a single individual have common goals; their behavior can be controlled by a single decision center; and therefore one may expect them to be much more strongly integrated than members of populations or communities. Thus, from the point of view of contemporary evolutionary theory, ecological analogies between cells, individuals, and populations are not justified.

In the study of ecological systems, the concept of the superorganism has to be rejected, and not only for theoretical reasons. The history of ecology seems to confirm the theoretical supposition that this concept does not generate testable scientific hypotheses. In my opinion, some old theoretical models of the predator-prey system that are aimed at predicting the behavior of ecological systems from the simple properties of individuals are still useful and important, whereas descriptions of communities based on analogies with individual organisms appear to be of little value.

Does this mean that the mathematical portion of ecological theory is free of ideas derived from the concept of the superorganism? When contemplating the basic ecological model of limited growth, the logistic equation (section 1.3), one sees that in fact individuals and their properties are not included in this model. It seems that mathematical ecologists, as well as other ecologists, were tempted by the holistic approach in their belief that a disregard for the properties of individuals is not a real obstacle to understanding ecological systems.

In order to understand how a given organism works, a biologist attempts to identify its elements, the elements' properties, and the relations among the elements. This reductionist approach seems to be the most efficient scientific method. If some biologists do not apply it, it is because they find identifying the system's elements and their properties either very difficult or simply impossible. Ecologists are in a completely different position: they usually know more about the elements of a system than about the system itself. With a few exceptions, it is easy to distinguish the individuals that are the elements of an ecological system and to identify their properties, but the entire system is much less integrated, and more difficult to define and to study. Ecological holism unfortunately ignores the knowledge already acquired about individuals and allows ecology to lose touch with the real world.

The reductionist method in ecology must derive the properties of ecological systems from the properties of their elements, i.e. individual organisms. To do this, the appropriate hypotheses and their more formal counterparts, mathematical models, are applied. There is a limit to the number of properties and factors that can be taken into account in a model; therefore, some factors and properties are assumed to be fundamental, while others are assumed to be random interferences of minor importance. We are also inclined to assume that all factors that are difficult to describe mathematically are random interferences that do not alter the basic prediction of the model.

Intrapopulation variability is the basis of the most general biological theory—the theory of natural selection—but this variability complicates enormously any mathematical description of ecological processes. Thus it is not surprising that current ecological theories usually ignore individual variation other than that due to sex, age, and some qualitative genetic differences. This book is an attempt to show how individual variability can be incorporated into the model of population dynamics and what the consequences of such an inclusion are. In this introductory chapter, the basic models of unlimited and limited growth are briefly discussed (section 1.2 and section 1.3, respectively), and then an alternative derivation of the logistic equation is given (section 1.4).

1.2. UNLIMITED POPULATION GROWTH

The model of unlimited population growth in discrete time, for nonoverlapping generations and generation time equal to the time unit, is given by the equation

$$\mathcal{N}(t) = \mathcal{N}(0)R^t, \tag{1.1}$$

where $\mathcal{N}(t)$ is the population size at time t, and R is a constant parameter called the net reproductive rate. This model in its continuous form, for both overlapping and nonoverlapping generations, is given by

$$d\mathcal{N}/dt = r\mathcal{N}, \tag{1.2}$$

where r denotes the intrinsic rate of natural increase.

The model of unlimited growth has a much wider application than simply to populations of separate organisms: it can be used to represent strictly physical or chemical processes or the growth of biological tissues, as well as the growth of an entire single organism. A population of separate organisms can also be viewed as a homogeneous substance, and the fact that it is made of individuals can be ignored. When growth is unlimited, it often does not matter whether a population is treated as a homogeneous substance or as a set of individuals. But this does matter when resources are limited, as I will try to show in Chapter 2; for this reason I will discuss later only models that explicitly consider individuals and that take into account differences in their individual properties.

The model of unlimited growth is the oldest and most general model in theoretical ecology, and it can be regarded as well established in ecological theory. Its generality stems from the phenomenon that all organisms originate by reproduction from other organisms, and therefore the number of organisms in the next generation is assumed to be related to the number in the previous generation. This model is based on individual properties of plants and animals, namely, the average number of progeny and the probability of survival.

It is an open question whether the probability of survival may be regarded as an individual's property. Since it can only be estimated as the proportion of survivors within a population, it is not strictly an individual feature. The point I would like to make is that although we cannot determine this probability from the fate of a single individual, we may conduct a separate experiment outside the studied population in order to estimate it. Therefore, the probability of survival may be regarded as an individual feature, like other features that are subject to independent estimation.

In a population with nonoverlapping generations, the net reproductive rate R can be defined as a product of the probability of survival from birth to reproduction, S, and the number of offspring per individual, B, namely,

$$R = SB. \qquad (1.3)$$

This equation predicts the expected value of the net reproductive rate R of a single individual or of a group of identical individuals, but if the parameters S and B are mean values of random variables, and there is a correlation between them, the mean value of R cannot be calculated as their product. Consider, for example, three individuals whose probabilities of survival are

$$S_1 = 0.25, \; S_2 = 0.50, \; S_3 = 0.75,$$

and whose offspring number

$$B_1, = 1, \; B_2, = 2, \; B_3, = 3.$$

The arithmetic mean of the probabilities of survival equals 0.5; these individuals produce two offspring on average; and the product of these two values equals 1.0. The net reproductive rate calculated in this way is not equal to the arithmetic mean of the three reproductive rates characteristic of the three considered individuals. The reproductive rates are:

$$R_1 = 0.25, \; R_2 = 1, \text{ and } R_3 = 2.25,$$

and their arithmetic mean equals 1.17. Weak individuals usually have both a low rate of reproduction and a low rate of survival, while strong individuals have high reproduction and survival rates; therefore, it is necessary to consider this correlation when calculating the net reproductive rate from the proportion of surviving individuals and the mean clutch size.

The variation in the net reproductive rate R among individuals does not change the model's prediction of population growth in future generations, unless this variation is hereditary. Thus, R can be applied both as a single value and as an arithmetic mean. This is obvious, because the arithmetic mean of R for N individuals, multiplied by N, equals the sum of the reproductive rates, which in turn equals the number of progeny produced by these N individuals. If there is a hereditary variation in R, natural selection will obviously increase the value of R in future generations, yielding a higher population size $N(t)$ than predicted.

The value of the model of unlimited population growth can be questioned if it is used to describe unlimited population decline ($r < 0$ and $R < 1$). In discrete time units, we can well imagine a situation where the product of the survival probability, S, and the number of progeny produced, B, is lower than unity, and at the same time constant for several generations; under such conditions the model can still be applied. On the other hand, in continuous time, if there is no reproduction and only deaths occur, there is no good reason to believe that population dynamics are determined by the equation of unlimited growth with $r < 0$. We may expect any one of several kinds of population decline: an immediate extinction of the entire population, an immediate extinction after a lapse of time, a straight line decrease, or an exponential decrease as predicted by the model given by equations (1.1) and (1.2).

It is well known from the statistical practice of the analysis of quantal responses (Hewlett and Plackett 1979) that the proportion of individuals that die due to a given concentration of an adverse factor often represents a positively skewed distribution in relation to this concentration and can be approximated quite well by the logarithmic normal distribution. Under detrimental conditions, time can be regarded as a concentration of an adverse factor, implying that large numbers of individuals die at first, and then the number of deaths diminishes, so that strong individuals survive for a long time. When survival is plotted as a function of time (Figure 1.1), it often appears that the same proportion of individuals dies in each time unit, and consequently the negative exponential function gives a good approximation of this process. There is no good reason why the same proportion of population members should die in every time unit, but there is good reason why the process is determined

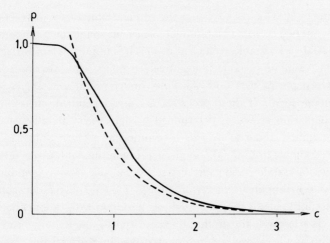

FIGURE 1.1. Hypothetical proportion p of surviving individuals as a function of the concentration c of an adverse factor, expressed by the exponential function given by $p = 2.43 \exp(-1.82c)$ (broken line) and by a function derived from the logarithmic normal distribution (solid line) according to the equation $p = 1 - F(\ln c)$, where F is the integral of normal distribution with the mean value of $\ln c = 0$ and its standard deviation $s = 0.5$.

by a skewed distribution of individual susceptibility to adverse conditions.

One can argue that the skewed distribution of individual susceptibility is yet another statistical epiphenomenon, providing no explanation of the origin of the distribution. This is not the case. In Chapter 3, I will try to show that the logarithmic normal distribution, and other kinds of positively skewed distributions of some important individual features, are general phenomena that can be derived from the individual properties of plants and animals.

The application of the equation of unlimited growth in its continuous form presents another problem. The equation can be regarded as an approximation of discrete growth for short time intervals and small changes of population size, but the main task of this model is to describe continuous birth and death

9

processes. This requires that the survival time or the time between two consecutive reproductive events should be a random variable. In simple ecological models it is usually assumed that this random variable is negatively exponentially distributed, which means that both mortality and reproduction are age-independent. If these processes are age-dependent, then the population growth is determined by age-specific natality and mortality, and also by age distribution at a given time. It is well known from the theory of A. J. Lotka that with given age-specific natality and mortality schedules, one obtains a stable age distribution after a certain time period, and only under such circumstances is the unlimited population growth exponential.

It is less well known that the same applies to microorganisms with simple life cycles. Continuous exponential growth describes growth by cell division if each cell has the same probability of fission in a unit of time. This requires a negative exponential distribution of time intervals between consecutive cell divisions, which implies that cells have no age, so that they have the same probability of fission immediately after a division or after any time lapse. This is obviously not how the microorganisms reproduce: they go through a cell cycle between divisions, which makes immediate fission impossible. Since the time intervals between divisions are not exponentially distributed, the theory of branching processes (Jagers 1975) must be applied to describe population growth at early stages, until the population approaches a stable age distribution. This limitation is not the result of the existence of a complicated life cycle, but rather of a much more general individual property of all organisms—the age-dependence of their natality and mortality. This is an important point: it shows that populations of both unicellular and multicellular organisms are not shapeless masses whose growth can be easily described by simple differential equations. Such equations can be used only as an approximate description.

When assuming nonoverlapping generations and discrete time, the model of exponential unlimited growth seems to be well established. Its applicability to populations with overlapping generations or in continuous time, however, is based on the assumption of stable age distribution, which must also be assumed for precise predictions of the population dynamics of unicellular organisms. The model's applicability to the description of population decline can also be questioned.

1.3. LIMITED POPULATION GROWTH

The most commonly used model for a population with limited growth is a well-known logistic equation, which in its continuous form is given by

$$dN/dt = r(1 - N/K)N, \tag{1.4}$$

where K is called either a point of equilibrium or the carrying capacity of the habitat, measured by the number of individuals that can be supported by this habitat.

Taking into account that population growth in a given moment of time is not determined by the number N of individuals at that same moment, a time lag can be introduced, or we may apply this equation in its discrete form (May 1972), namely,

$$N(t+1) = N(t) + N(t)(R-1)[1 - N(t)/K]. \tag{1.5}$$

Equations (1.4) and (1.5) are very simple and elegant mathematical structures, but it does not seem likely that anyone would accept them as a formal description of a testable hypothesis. It is unreasonable to predict that if $N = 2K$, then the rate of population decrease is equal to $-r$. Similarly, we know many species of plants and animals with discrete generations that produce tens or hundreds of offspring in the same unit of time. Without the adverse effect of intraspecific competition, most of these offspring are able to survive to the time of their own reproduction; this implies that the number of female offspring is equal to the reproductive rate R, so that this rate equals tens or hundreds of females born for every female. If so, then according to the stability criterion for the discrete model (May 1977), which predicts local stability only if $R \leq 3$ and cycles if $R \leq 3.57$, one can expect such populations to be locally unstable and subject to frequent extinctions.

It does not therefore, make sense to expect that equations (1.4) and (1.5) predict the dynamics of natural populations. Even if the logistic theory did predict correctly the growth of populations, this would not change its present status, because we require both the assumptions and the predictions of hypotheses to be consistent with reality. The criticism raised below concerns the heuristic value of the logistic equation as a theoretical structure aimed at clarifying some of the aspects of limited population growth.

The first shortcoming of the logistic equation is that its assumptions are not based on biological knowledge of plants, animals, and microorganisms. Imagine an organism that is able to perceive its population density N in relation to the number K of individuals that can be safely supported by the available resources, and imagine further that the organism reduces its reproductive output proportionally to the N/K ratio. We do not know of such an organism, although it is evolutionarily possible if all population members belong to the same clone and are genetically identical. Such a reduction in reproductive capacity would be advantageous for each individual if the depletion of resources were a real and imminent danger. But even under these special circumstances, the logistic equation would be of little value. To understand such a system of self-regulation within a clone, the model should be based on a quantitative description of an individual's reactions to the presence of others, in order to predict the behavior of the entire clone from the interrelations among its members. Only if its members are assembled as a single colony such that their identification and the study of their individual behavior is difficult will the epiphenomenological description make sense.

The logistic equation may be seen from the point of view of the properties of the population members, if we rearrange equation (1.4) to obtain

$$(1/N)(dN/dt) = r(1 - N/K), \qquad (1.6)$$

which is the per capita population growth, as a decreasing function of population density. Is such per capita growth the property of an individual? If it includes the birth rate only, it obviously is. We can measure the number of offspring produced by an individual, whereas (as mentioned in section 1.2) the probability of survival can be measured only for a group of individuals, since a single individual can only either die or survive. On the other hand, by studying the survival of a group of organisms that differ with respect to age, weight, and physiological condition, we can discover how these properties influence their probability of survival. On this basis we can assign a probability of survival to a single individual in a population, and by including birth rate we can consequently assign a reproductive rate to a single individual.

We now must ask whether the logistic equation assumes an identical response to crowding from all population members, or whether it allows for individual variation. According to Poole (1974) and Pielou (1977), crowding affects all individuals equally, so that the individual reproductive rates lie at the line determined by the logistic assumption. On the other hand, the logistic assumption is often supported by empirical data, like those presented in Figure 1.2 in the form of a scatter diagram of decreasing individual birth rate or of decreasing viability with increasing population density. The data presented in Figure 1.2, taken from Rubenstein (1981a), have not actually been used by him to support the logistic equation, but they show clearly how the birth rate changes with increasing density. One can argue that the basic feature of limited population growth is given by the regression line (it can be either a logistic straight line or some other line based on a modified assumption), while the variability in the scatter diagram presents random deviations, common to all biological measurements, that have to be

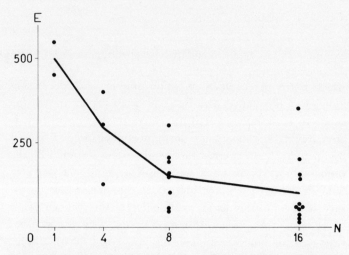

FIGURE 1.2. Number of eggs produced by individual females of Everglades pygmy sunfish (E), at different densities (N). Each point represents the progeny of one female. Solid lines connect the points of arithmetic means for each density. Modified from Rubenstein (1981a).

ignored as unimportant random variation in order to explain the basic features of the system.

Is intrapopulation variation really an unimportant feature of the system? If resources become scarce, population members refrain from reproduction and die of hunger. If members are all identical, they die simultaneously and, as mentioned in section 1.2, there is no reason to believe that the decrease of the population of identical individuals will be exponential. But something quite different can be expected when large and well-pronounced differences among population members exist: some individuals will die promptly, leaving space and resources to others who will survive and even reproduce. Such behavior can make the population very stable. This phenomenon will be discussed in detail in Chapter 2, but even without close examination, it seems obvious that individual variability is crucial to population stability. For example, territorial behavior among animals brings about stability because it allows only part of a population to breed.

Since the logistic model ignores individual variability, both in its classical form and in its various modifications, it ignores an important source of population stability. This seems to be the most important shortcoming of the logistic model.

However, one should not completely deny the usefulness of the logistic equation. Theoretical ecologists have used it to show the meaning of the term "stability" and the effects of time lag, in both discrete and continuous time. The equation was obviously a source of inspiration for many laboratory and field ecologists. On the other hand, because it ignores individual population members, the logistic equation has not described the mechanism of intraspecific competition. Since the equation is an important part of ecological theory, many ecologists have thus neglected this kind of competition. Consequently, ecological textbooks concentrate more on interspecific than on intraspecific competition; about the latter we can learn more from books on animal behavior or natural history. The phenomenon of competition among the members of the same species, which is of fundamental importance both for ecology and the theory of evolution, has been lost in the logistic equation, which says that population size increases if $N < K$ and decreases if $N > K$, without considering any specific mechanism of the process.

Some attempts to construct more realistic models of limited population growth, starting from the biological properties of organisms, have been made, for example, by Stewart and Levin (1973) and Schoener (1973); but since these models also have ignored the differences between individuals, I will not discuss them here.

1.4. LIMITED GROWTH IN DISCRETE PLACES

It is possible to derive the logistic equation from the properties of individual members of a population. Consider a patchy environment, such that each patch is a place in which an individual organism can survive and leave progeny. In a unit of time an individual either dies with probability E without leaving progeny or reproduces giving R offspring and then dies. (Note that the exact meaning of R here differs from that in the previous section, where it denotes the net reproductive rate. Here it denotes simply the number of offspring.) Only one of R offspring can stay in this place, while $(R-1)$ must emigrate to find other places to live. Among the emigrating individuals only a fraction M survive to reach another favorable spot. If there is a space with H places that can be occupied, and the proportion p of these places is currently occupied, then at each time unit there are $pH(R-1)M/H = p(1-R)M$ colonizers, per place. If these colonizers are randomly distributed, then from the Poisson distribution the proportion of places into which at least one immigrant arrives is given by

$$C = 1 - \exp[-p(R-1)M], \qquad (1.7)$$

which can be regarded as a probability of colonization of an empty place, in the same way that E is the probability that the place previously occupied is now empty. The proportion of the occupied places in the next time unit, $p(t+1)$, is given by

$$p(t+1) = p(t) + [1-p(t)]C - p(t)E. \qquad (1.8)$$

Since $C = 1 - \exp(-pU)$, where $U = (R-1)M$, no analytical solution for equation (1.8) can be found, and we must simplify the system further by removing p from the exponent. Therefore, we define the probability of colonization as

$$C = pU, \qquad (1.9)$$

which is a good approximation of equation (1.7) if $0 < p < 0.1$. This approximation implies a further constraint upon the model's applicability, since a very low density of colonizers per place, pU, is required. This constraint can always be added by shortening the time unit, but then the model will no longer have a time unit equal to generation time.

Substituting equation (1.9) into equation (1.8) yields

$$p(t+1) = p(t) + [1 - p(t)]p(t)U - p(t)E. \qquad (1.10)$$

The identical model in continuous time was applied by Levins (1970) for the dynamics of a metapopulation, designed for the study of the mechanism of group selection. Levins defined U as a colonization rate. I prefer to call it colonization ability, determined by individual reproduction and survival, while the colonization rate, pU, is additionally dependent on the proportion of places occupied.

The number of individuals that are established in their places and able to reproduce is given by

$$N(t) = p(t)H. \qquad (1.11)$$

Multiplying equation (1.10) by H we obtain

$$N(t+1) = N(t) + N(t)[1 - N(t)/H]U - N(t)E. \qquad (1.12)$$

The nontrivial equilibrium point K at which $N(t+1) = N(t) = K$ is given by

$$K = H(1 - E/U). \qquad (1.13)$$

Similar results were obtained by Kozłowski (1980) in a continuous model, when he attempted to derive the logistic equation from the Volterra and Lotka predator-prey equations by assuming that birth rate is proportional to the fraction of space unoccupied. Substituting equation (1.13) into (1.12), we obtain

$$N(t+1) = N(t) + (U - E)N(t)[1 - N(t)/K], \qquad (1.14)$$

which is obviously the logistic equation as presented by equation (1.5), but with different meanings for the parameters. K is the equilibrium point, but with the death rate E higher than zero, it is always smaller than the number H of available places. Since K is determined by the death rate E and by recruitment, as described by colonization ability U, it can hardly be called the carrying capacity. We should rather say that, according to equation (1.13), K is an equilibrium point that is linearly related to carrying capacity H—in other words, to the number of available places. The parameter U can be regarded as a birth rate, but it also includes mortality at the migratory stage, before competition occurs. Thus, the term $(U - E)$ is equivalent to an intrinsic rate of natural increase when intraspecific competition is absent. If $E > U$, the population size decreases; similarly, the population decreases if $r < 0$. The latter property is also valid when equation (1.7) is applied instead of (1.9), as shown elsewhere (Lomnicki 1980a).

How stable is this model of population regulation? From equation (1.8),

$$[dp(t + 1)]/[dp(t)] = 1 - C - E.$$

Since both E and C are probabilities, they cannot exceed unity, and so the system is stable for all sets of parameters, according to the stability criterion for difference equations: $|dN(t + 1)/dN(t)| < 1$. It is impossible to overshoot resources, because there are only H places in the considered space; individuals that are not in these places do not participate in reproduction and therefore are not counted as population members.

From the biological point of view, the stability of the system results from inequality among population members: those lucky enough to have places to live and to reproduce cannot be displaced by newly arrived immigrants. A linear decrease in the rate of recruitment with increasing density results from random dispersion of individuals in the space considered.

It is difficult to see how such a model can be applied to resources other than a set of distinct favorable places, each of which is able to support one individual or a family. When resources are in discrete units, some individuals can take a unit while others get nothing. This brings about individual variability, which in turn will stabilize population size. If resources are not discrete, a quite different model is required that considers how the resources are divided (Chapter 2).

Another approach to the study of the stability of populations distributed among small patches was proposed by de Jong (1979) and developed by Hassell and May (1985). The stability in their model is based on the clumping of individuals among patches. This model is discussed in section 10.3.

CHAPTER TWO

Individual Variation in Resource Partitioning and Population Dynamics

2.1. RESOURCE PARTITIONING AMONG INDIVIDUALS

In the ecological literature, a great deal has been written on resource partitioning among different species within a community, but data on resource partitioning within a single-species population are scarce. Laboratory studies can measure the nutrients that an individual plant or a group of plants is able to take, requires, or actually takes. It is also relatively easy to determine the food intake of a single animal or a group of animals. However, it is technically very difficult to evaluate the amount of nutrients or of food taken by a single individual within a group. This may well be the reason why most studies on energy and nutrient flow through ecosystems have been based on measurements either of separate individuals or of whole groups, and why studies have usually ignored the variation of these measurements among individuals. Nevertheless, we do know that individuals within a single population differ in size and weight, and that three important bioenergetic processes (assimilation, respiration, and production) are strongly weight-dependent. Thus, we can expect large differences in the nutrient and food intake among individuals within a single population.

In Chapter 3 a more detailed discussion on the origin and pattern of individual variation is given. The present chapter

examines the effects of individual variation in resource partitioning on population stability. A simple model of population
dynamics is proposed, with four versions of resource partitioning.
To facilitate the analysis, I have restricted the model of population dynamics to the case of nonoverlapping generations
with the time unit equal to the generation time. I have also
assumed the same resource inflow, V, to the population in each
generation. These simplifications make analytical solutions possible, which in turn help to reveal the effects of individual
variability on the behavior of the entire population. A simple
extension of the model to overlapping generations is given in
section 5.3; the justification for the use of discrete models instead
of continuous ones is given in section 5.4.

Differences in resource partitioning, like any other differences,
can be described by the distribution of individual resource intake; but in the models presented here, it is much more con-
venient to rank individuals from the first, with the highest intake
of resources, to the last, with the smallest intake. Such a ranking
allows for the interpretation of an individual's resource intake
y as a function of its rank x. If two or more individuals take the
same amount of resources, their ranks are a matter of convention.
Assigning the rank x does not imply the existence of linear social
hierarchy, but if such a hierarchy exists, resource intake y and
rank x may well be correlated with it.

The data on weight distributions discussed in Chapter 3 suggest that distributions of the individual resource intakes are
positively skewed, and are often close to logarithmic normal
distributions, as shown in Figure 2.1(A). From this distribution,
as from any other, the function $y(x)$ (Figure 2.1[C]) can be
derived by calculating the integral of this distribution $F(y)$
(Figure 2.1[B]). The inverse of the function $[1 - F(y)]$, with
the horizontal axis multiplied by the number of individuals N,
is equivalent to the function $y(x)$.

An individual's rank x can also be interpreted as the number
of individuals that have a resource intake higher than or equal
to $y(x)$. Another approach to describing unequal resource par-

21

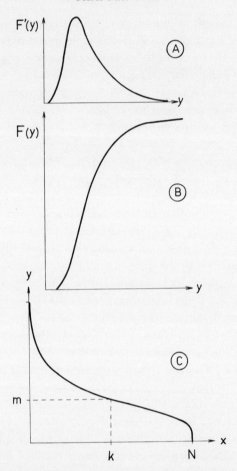

FIGURE 2.1. The relation between the distribution of individual resource shares y (A), its cumulative distribution (B), and the function $y(x)$ (C). This is an example of log-normal distribution. On the diagram of the function $y(x)$, the number k of individuals that obtain at least m resources out of all N individuals is shown.

titioning is to apply not the rank x, but a proportion $p(y)$ of individuals that take at least y units of resources. The product of this proportion $p(y)$ and the size N of a population is equal to the rank x of an individual that takes exactly y units of resources. This last interpretation clearly shows that a linear social hierarchy is not required here. However, the application of proportion $p(y)$ instead of rank would make further analysis more involved, and therefore, for simplicity, the concept of rank x has been applied here.

The general model of population dynamics is as follows. Every individual can take no more than a units of resources, so that $y(x) \leq a$, and an individual requires at least m resource units to survive to the reproductive period. Therefore, a can be interpreted as the satiation level of an individual, m as its maintenance cost. Individuals that take fewer resources than the maintenance cost—in other words, those for which $y(x) < m$ — die before reproducing. The number of individuals k (Figure 2.1 [C]) that receive more than m units of resources is given as the rank of an individual that receives exactly m units of resources, according to the general equation

$$m = y(k). \tag{2.1}$$

If we assume that those resources not used for maintenance are allotted to reproduction, the amount of resources $g(x)$ allotted to reproduction by an individual of rank $x \leq k$ is given by the equation

$$g(x) = \max\{0, [y(x) - m]\}. \tag{2.2}$$

The number of individuals in the next generation $N(t + 1)$ is determined by the sum of resources $g(x)$ allotted to reproduction by individuals of the present generation, according to the equation

$$N(t + 1) = h\sum g(x), \tag{2.3}$$

where h denotes the efficiency of converting resources into offspring, and $\sum$ denotes the sum over all the k individuals. The fundamental property of this model as presented by the equations given above is that the individuals can really obtain unequal resource shares. For example, if $y(50) - m = 20$ and

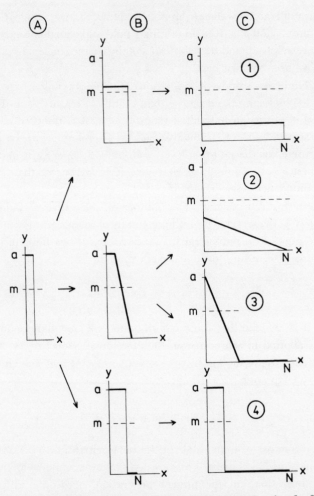

FIGURE 2.2. Individual resource intake y as the function of rank x for four different versions of the model described in the text. Three stages (A, B, C) result from increasing population size N. Note that $k(N)$ presented in Figure 2.3 are the values of x for $y = m$, while $N(t+1)$ presented in Figure 2.4 are linearly related to the areas delimited by vertical axes, $y(x)$, and the line parallel to the horizontal axes at point m.

$y(100) - m = -5$, the shortage of nutrients or food for the 100th individual is not compensated by the large amount of these resources for the 50th individual. This phenomenon is ignored in most ecological models, which consider the global intake of resources and the global production of the entire population, without regard to individual resource intakes and the fates of individuals within the population. In the example mentioned above, an individual with rank $x = 50$ gives $20h$ offspring, while the one with rank $x = 100$ dies, leaving no progeny at all.

One can expect that individuals with a higher resource share y may have a higher maintenance cost m, and that they may produce offspring of higher quality, but these phenomena are omitted in the model presented here, for the sake of simplicity.

2.2. FOUR VERSIONS OF THE MODEL OF RESOURCE PARTITIONING

The versions of the model of resource partitioning among individuals presented here are conceptually simple, but require numerous equations for their description. Should the equations appear too complicated, the diagrams representing them (Figure 2.2) will reveal the model's simplicity. The function $y(x)$ is represented in each version by straight lines; this is not very realistic, but it makes simple analytical solutions possible. Because of this simplification, one cannot expect precise quantitative predictions concerning population dynamics, only some qualitative ones.

As shown below, the four versions of resource partitioning presented on Figure 2.2 determine the number of individuals k that survive to the time of reproduction (Figure 2.3) and the number of individuals $N(t+1)$ in the next generation (Figure 2.4), as functions of $N(t)$. For all four versions I have assumed that a population of $N(t)$ individuals has V units of resources during a unit of time (one generation). At low population density, each individual takes as many resources as it requires,

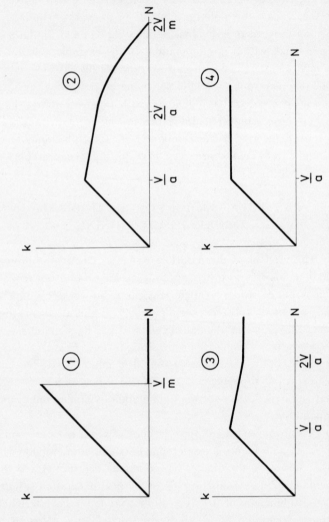

FIGURE 2.3. Number of individuals that survive to reproduction k as a function of the initial number of individuals $N = N(t)$ for the four versions of the model presented in Figure 2.2. V denotes the total amount of resources for the entire population, while a is the maximum value of y.

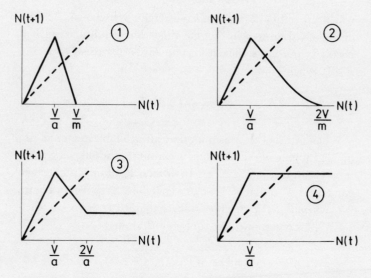

FIGURE 2.4. $N(t+1)$ as a function of $N(t)$ for the four versions of the model presented in Figure 2.2. V denotes the total amount of resources for the entire population, while a is the maximum value of y.

namely, a units. This means that for $N(t) \leq V/a$—in other words, for a population size so small that each individual can take a units—

$$y(x) = a, \qquad (2.4)$$

and all $N(t)$ individuals survive to reproduction, so that

$$k = N(t). \qquad (2.5)$$

It follows from equation 2.2 that each individual uses $(a - m)$ units of resources for reproductioin, and the population size in the next generation is given by

$$N(t+1) = (a-m)hN(t), \qquad (2.6)$$

27

which, by setting $R = (a - m)h$, yields the well-known equation of unlimited growth in discrete time. The condition which is necessary but not sufficient for population persistence requires $R \geq 1$, which implies that

$$(a - m)h \geq 1. \tag{2.7}$$

When $N(t) > V/a$, I assume that all available resources are divided among $N(t)$ individuals according to the following rules, different in the four versions. In version 1, all individuals take equal shares, so that $y(x) = V/N$. Therefore, as shown in Figure 2.2, when $V/a < N(t) < V/m$, and consequently $m < y < a$, then $k = N(t)$ as determined by equation (2.5), and

$$N(t + 1) = [V/N(t) - m]hN(t). \tag{2.8}$$

If $y(x) < m$, all individuals must die before reproducing; therefore, for $N(t) > V/m$,

$$k = 0, \tag{2.9}$$

and

$$N(t + 1) = 0. \tag{2.10}$$

In versions 2 and 3, $(2V/a - N)$ individuals take a units of resources, while others get a smaller share that decreases linearly with their rank x, so that all resources are used; in other words, the area under the $y(x)$ function is equal to V. It follows from elementary geometry that for $V/a < N(t) < 2V/a$,

$$k = N(t) - 2(m/a)[N(t) - V/a], \tag{2.11}$$

and

$$N(t + 1) = h[V(1 - m^2/a^2) - mN(t)(1 - m/a)]. \tag{2.12}$$

With a further increase in population size $N(t)$, and observing the convention that individual resource share is a linearly decreasing function of individual rank x, there are two possible ways to partition resources, as represented by versions 2 and 3. In version 2, the shares of high-ranked individuals (those of low x values) are reduced below the maximum possible level a; with increasing $N(t)$ they are further reduced, even below the minimum required m; and these resources are distributed among all population members. In version 3, the high-ranked individuals take almost the maximum possible amount of resources, so that next to nothing or nothing is left to low-ranked population members. Note that in version 3, the presence of individuals of the lowest ranks does not affect the resource intake of those of higher ranks. From the rules outlined above and presented in Figure 2.2, version 2 yields

$$k = N(t) - m[N(t)]^2/(2V), \qquad (2.13)$$

and

$$N(t + 1) = h\{V - mN(t)[1 - mN(t)/(4V)]\}, \quad (2.14)$$

for $2V/a < N(t) < 2V/m$, whereas for $N(t) \geq 2V/m$, equations (2.9) and (2.10) hold. In version 3,

$$k = 2(V/a)(1 - m/a), \qquad (2.15)$$

and

$$N(t + 1) = hV(1 - m/a)^2, \qquad (2.16)$$

for all values of $N(t) > 2V/a$, which implies that both k and $N(t+1)$ are independent of $N(t)$ in this range of population density.

Version 4 represents the extreme case of unequal resource partitioning. When the population size $N(t)$ is larger than V/a, then the V/a individuals of the highest ranks take a units of

29

resources each, while others get nothing. This yields $k = V/a$ as given by equation (2.5), while

$$\mathcal{N}(t+1) = hV(1 - m/a), \qquad (2.17)$$

which implies that $\mathcal{N}(t+1)$ is independent of $\mathcal{N}(t)$.

Even without following equations (2.4) to (2.17), Figures 2.2, 2.3, and 2.4 allow us to deduce the different ways resources can be partitioned among members of a population and the consequences of these differences for the survival of individuals and for population dynamics. We see a gradient from equal partitioning, where the share $y(x)$ of all individuals is dependent on the population size $\mathcal{N}(t)$, to unequal partitioning where shares for high-ranked individuals are independent of population size. In the former case, $\mathcal{N}(t+1)$ is a function of $\mathcal{N}(t)$; in the latter, it is independent of $\mathcal{N}(t)$ for $\mathcal{N}(t) \geq V/a$.

2.3. POPULATION STABILITY AND PERSISTENCE

Stability and persistence of populations with resource partitioning as described by the model presented above can be studied according to the rules of the stability analysis of difference equations (May and Oster 1976; for a popular account, see Vandermeer 1981). An example of such an analysis is given in Figure 2.5. The equilibrium point $\mathcal{N}_e$ lies at the intersection of the line representing the function $\mathcal{N}(t+1) = f[\mathcal{N}(t)]$ and the broken line. This broken line is a diagonal that serves to plot the number of individuals from one axis to another. The equilibrium point is locally stable if the slope of $\mathcal{N}(t+1)$ as the function of $\mathcal{N}(t)$ at the equilibrium is smaller than unity (smaller than 45 degrees)—in other words, if

$$|[d\mathcal{N}(t+1)]/[d\mathcal{N}(t)]| < 1. \qquad (2.18)$$

The population can persist if it is able to recover from low density—that is, if the slope of the increasing part of $\mathcal{N}(t+1)$

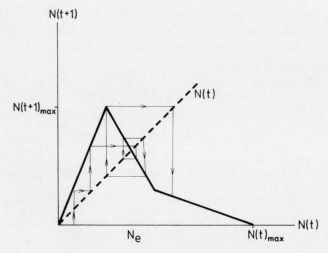

FIGURE 2.5. An example of the stability analysis of difference equations, applying the graph of $N(t + 1)$ as a function of $N(t)$. Starting from any point on the $N(t)$ axis, one can determine population size in the next generation. This newly determined population size $N(t + 1)$ can be moved on the $N(t)$ axis by using the broken line, which has a slope of 45 degrees. By consecutive plotting from the broken to the solid line one can determine the dynamics of the system. The equilibrium point is at the intersection of the solid line and the broken one. Note that the equilibrium point is unstable on this graph, because at this point the slope of the solid line [$N(t + 1)$ as the function of $N(t)$] is higher than unity. Nevertheless, this system is persistent, since $N(t + 1)_{max} < N(t)_{max}$, and therefore the present population cannot attain the size above which the next generation's population will drop to zero.

as the function of $N(t)$ is higher than unity, so that condition (2.7) is fulfilled—and if

$$N(t + 1)_{max} < N(t)_{max}, \qquad (2.19)$$

where $N(t)_{max}$ and $N(t + 1)_{max}$ are maximum values of $N(t)$ and $N(t + 1)$, respectively, as presented in Figure 2.5. Condition (2.19) means simply that if $N(t + 1)_{max}$ is higher than $N(t)_{max}$, then the population size may drop to zero in the next generation. If the maximum value of $N(t + 1)$ is smaller than the maximum value of $N(t)$, the population will never attain a size that allows complete extinction to occur. Note that a population can be persistent without being locally stable, just as it can be locally stable but not persistent, if population size is removed from the point of local stability.

When applied to version 1 (Figure 2.4 [1]), condition (2.18) for local stability yields

$$1/h > m. \tag{2.20}$$

Since $\mathcal{N}(t)_{max} = V/m$ and $\mathcal{N}(t+1)_{max} = hV(1 - m/a)$, condition (2.19) for persistence is given by

$$1/h > m(1 - m/a). \tag{2.21}$$

Condition (2.20) carries a biologically important message. Since h denotes the efficiency of converting resources into progeny, therefore $1/h$ is the cost of producing a single offspring. This implies that for the population to be locally stable, the maintenance cost m has to be smaller than the cost of producing a single offspring. It seems that for the majority of plants and animals, the cost of maintenance is much higher than the cost of producing one offspring. This is certainly the case for organisms that produce hundreds or thousands of propagulae, but even for those which produce few offspring, the cost of reproduction per single offspring is much lower than the parents' maintenance cost.

Condition (2.21) for persistence is slightly less restrictive but is still difficult to meet. If these two rigid conditions are not met, a population in which resources are equally divided cannot exist. Wilson (1980, p. 51) reached a similar conclusion for a special numerical example of equal resource partitioning among population members.

In version 2 (Figure 2.4 [2]), local stability depends on the position of the equilibrium population size $\mathcal{N}_e$ on the curve representing $\mathcal{N}(t+1)$ as the function of $\mathcal{N}(t)$. If it is placed in the part of the $\mathcal{N}(t+1)$ curve determined by equation (2.12), then the condition for local stability is given by the inequality $1/h > m(1 - m/a)$, which is identical to the condition for persistence in version 1. For higher $\mathcal{N}(t)$ values, the slope of the $\mathcal{N}(t + 1)$ curve, as determined by equation (2.14), gradually decreases to the value of zero at the point where $\mathcal{N}(t)_{max} = 2V/m$, which is twice as high as $\mathcal{N}(t)_{max}$ for version 1. Therefore condition (2.19) for the persistence of the population is given by

$$1/h > m(1 - m/a)/2. \qquad (2.22)$$

This is essentially the same condition as for version 1, except that it is less restrictive, because the cost of producing one offspring can be half the cost in version 1. This condition for version 2 is only quantitatively different from version 1, with low stability and no buffering against extinction.

We obtain quite different results when analyzing version 3. In this version, in addition to unequal resource partitioning, it is assumed that an increase of population size $\mathcal{N}(t)$ above $\mathcal{N}(t) < 2V/a$ does not diminish the resource share $y(x)$ of high-ranked individuals. This in turn makes $\mathcal{N}(t + 1)$ independent of $\mathcal{N}(t)$, as mentioned above. Consequently, population extinction is impossible, because $\mathcal{N}(t + 1)$ does not drop to zero for high population densities $\mathcal{N}(t)$. If the equilibrium population size $\mathcal{N}_e$ lies below $2V/a$, there is still room for oscillations within the frame of this version of the model, but they cannot lead to extinction. The essential properties of this model—persistence and, depending on the set of parameters, the possibilities of oscillations—are the same as in a model of a host population regulated by microparasites presented by May (1985).

In version 4 the population is even more stable than in version 3. The resource partitioning postulated by version 4 brings about both local stability and population persistence for all sets of parameters, as long as condition (2.7), which allows population increase at low densities, is satisfied (Łomnicki 1980c). This is because $\mathcal{N}(t + 1)$ is either an increasing function of $\mathcal{N}(t)$ for

33

$\mathcal{N}(t) \leq V/a$ or independent of $\mathcal{N}(t)$ at its higher values. In version 4 population members use resources most efficiently, and this results in the highest population equilibrium density in relation to available resources (Figure 2.4 [4]).

Thus, it is clear from this analysis that population stability and persistence, in discrete time models, require not only unequal resource partitioning, but also the inaccessibility to those of lower ranks of resources controlled by high-ranked individuals. This can be guaranteed, for example, by the ability of some population members to guard their resources against other individuals of lower ranks. Such inaccessibility, which will be discussed more extensively in section 2.5 as well as in Chapter 6, seems to be even more important than the inequality of resource partitioning itself.

The relation between stability and persistence on one side, and inequality of resource partitioning and their monopolization on the other, seems to be a general one, and as shown in section 5.3, it also holds for overlapping generations. On the other hand, overlapping generations imply that there are several age groups within a local population, the existence of which can be an important source of intrapopulation inequality.

2.4. LABORATORY AND FIELD DATA

As mentioned earlier, it is rather difficult and in many cases impossible to determine empirically the function $y(x)$, describing resource partitioning among population members. On the other hand, it is relatively easy to find out how the population size in one generation determines the population size in the next generation or, in other words, to perform experiments that determine the empirical dependence of $\mathcal{N}(t+1)$ on $\mathcal{N}(t)$. Unfortunately, such data are also scarce. Usually, they do not span the entire life cycle of the organism studied to include its reproduction and the number of its progeny, but instead follow only a part of the life cycle during which the determination of the survival at different densities is made. A set of such data concerning insects can be found in the papers by Hassell (1975) and Bellows (1981). Since most of the data do not include reproduction, they can only show the empirical dependence of the number of survivors k as the function of the number N of individuals entering into competition for resources.

The data presented below are not necessarily the best available in the ecological literature. They were selected in order to illustrate some interesting points concerning intraspecific competition, as determined by resource partitioning among individuals.

To support version 1, some data collected by Harper (1961, cited in Harper 1977) on the measurements of the yield of maize ears sown at three different densities are applied here. The yield of the ears is given in kilograms, and I have assumed that there are 500 seeds per kilogram of ears. Using this estimate, the yield of seeds for these densities has been calculated and presented in Figure 2.6. The slope of the line $N(t + 1) = f[N(t)]$ is very steep, despite this rather low estimate of seed number per kilogram. Thus condition (2.18) for population stability is not fulfilled. These data do not tell us at which seed density the crop drops to zero. Such a density probably does not exist, since it is very difficult to imagine a density so high that not a single maize plant would be able to produce seeds. If so, then version 3 should be applied here. However, using the linear interpolation of the available data (Figure 2.6), we obtain the equilibrium density $N_e = 33$ seeds per square meter, and $N(t)_{max} = 40$ seeds. Since $N(t + 1)_{max}$ is estimated to be 200 seeds per square meter, the population of maize as described in figure 2.6 is unstable and may very easily become extinct.

On the basis of the three measurements presented in Figure 2.6, it is difficult to support the statement that the population of maize is really unstable and subject to frequent extinctions; these measurements suggest only that such a possibility exists. If there are any organisms that might suit version 1, they will

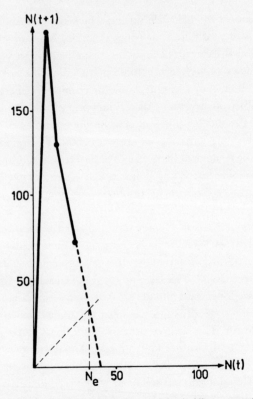

FIGURE 2.6. Maize yields, expressed as ear weights in kilograms multiplied by 500, $N(t+1)$, as the function of the number of seeds sown per square meter, $N(t)$. After Harper (1977, p. 197).

be cultivated plants. This is because during the process of plant cultivation there is very strong selection for uniform germination time, since seeds that delay germination are not able to leave progeny among the seeds collected to be sown by man. A synchronized germination results in the low variance of plant sizes and consequently in equal resource partitioning among individual plants, similar to that given by version 1. This is quite different from the situation among wild plants, where high variance in germination time and the ability to refrain from germination under unsuitable conditions (Harper 1977) generates individual variation. This in turn implies unequal resource partitioning and higher stability.

It is well known both in agriculture and in forestry that too-high densities result in lower yields. This does not mean that the yield can drop to zero as proposed by version 1, but that there is a wide range of densities $N(t)$ for which $N(t+1)$ is a decreasing function of $N(t)$. For this reason cultivation often requires seedlings to be cleared, whereas so-called self-thinning works well only in more variable natural populations.

The phenomenon discussed here was studied and analyzed by Nicholson (1954), who pointed out (p. 19), "The important part played by the wide scatter of the properties of animals and those of their environments in population dynamics cannot be overemphasized." The versions of the model presented here can serve as a formal description of Nicholson's concepts of scramble and contest competition. For example, he proposed that population size after competition is a decreasing function of the size before competition, because some food is consumed by those individuals that fail to mature. This is exactly what version 1, 2, and 3 show, when they allow for $0 < y(x) < m$. Individuals for which the above condition is fulfilled take food share $y(x)$, but this is used entirely for their maintenance, so they produce no progeny. This makes $N(t+1)$ a decreasing function of $N(t)$.

To illustrate the concept of scramble competition, Nicholson (1954) presented the well-known data on the survival of the larvae of the blowfly *Lucilla cuprina* (Figure 2.7). This classic example is used here as an empirical counterpart of version 2 of the model. Nicholson's data refer only to the survival of larvae; therefore, a theoretical counterpart of his data is given by the function $k(N)$ (Figure 2.3 [2]). Although the shapes of these two $k(N)$ lines differ, their basic features are the same: an optimum density at which the largest number of individual survive, and a decrease in the number of survivors above this optimum, which at a certain density reaches zero. The proposed shape of the $y(x)$ function described by the straight lines is

37

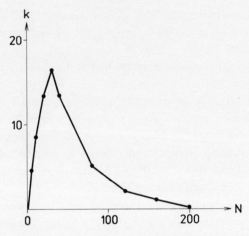

FIGURE 2.7. Number of *Lucilla cuprina* larvae surviving to metamorphosis, *k*, as the function of their initial density *N*. Modified from Nicholson (1954).

obviously too simplified to give a precise description of resource partitioning among the blowfly larvae, and consequently to predict the exact shape of $k(\mathcal{N})$. Nevertheless, since $k(\mathcal{N})$ decreases to zero, this suggests that resource partitioning among these larvae can lead to oscillations and frequent population extinctions. On the other hand, predictions based on the study of groups of larvae of similar age, as in Nicholson's experiments, are not sufficient to predict the dynamic of entire populations of these insects. Such laboratory populations, also described by Nicholson (1954), include larvae of various ages, imagines, eggs, and pupae. Since at present we are unable to describe such populations in terms of resource partitioning among individuals, we must rely on the models proposed for these populations by Readshaw and Cuff (1980) or Gurney et al. (1983). These models use the function $k(\mathcal{N})$ (Figure 2.7) from Nicholson's (1954) data, but they do not analyze the mechanisms that determine the shape of this function.

The survival of blowfly larvae is not the only example that can serve as an empirical counterpart of version 2. The number of surviving individuals of beetles of the species *Lasioderma serricorne* and *Tribolium castaneum* (Bellows 1981), as well as *Tribolium confusum* (section 5.2, Figure 5.3), as functions $k(\mathcal{N})$ of their densities $\mathcal{N}$, exhibit features similar to the theoretical $k(\mathcal{N})$ function of version 2.

The basic feature of version 3 is that at very high densities neither the number of survivors, k, nor the population size in the next generation, $\mathcal{N}(t+1)$, drops to zero, but both k and $\mathcal{N}(t+1)$ depend on $\mathcal{N}(t)$ at low densities. This can be explained by the following behavior of individuals: at low density, when rersources are abundant, additional individuals entering the population can diminish the resource intake y of individuals that are in a better position; but at high densities with a real shortage of resources, the individuals in better positions are able to monopolize their resources to the exclusion of others. Their resource intake y may be much smaller than at low densities, but these resources are defended from others.

Version 3 can be illustrated, for example, by the data on the survival of *Rana silvatica* tadpoles from the early developmental stage to metamorphosis (Wilbur 1976) presented in Figure 2.8. As in version 2, I do not claim that resources are divided between the tadpoles exactly according to the rules given in the respective

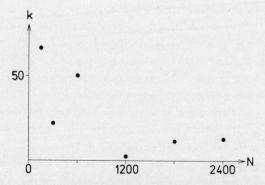

FIGURE 2.8. Number k of *Rana sylvatica* tadpoles surviving to metamorphosis as a function of their initial density $\mathcal{N}$. Data from Wilbur (1976).

39

versions (Figure 2.2). Since the larvae exhibit skewed weight distribution (Wilbur 1976), we can expect that there are more individuals with food intake $y < m$ than proposed by version 3 (Figure 2.2), so that more food is used by those which are later unable to metamorphose, and the group of those surviving at high densities is much smaller. This should make the hump in Figure 2.3 (3) more pronounced.

Of importance in version 3 of the model is the ability of high-ranked individuals to take enough resources, so that even at very high densities some larvae can metamorphose. Note that this ability was studied under experimental conditions in which the sizes of young larvae were initially as uniform as possible and the habitat in which competition took place was homogeneous. In spite of this, there were survivors even at the highest densities. Since some tadpoles are able to metamorphose at the highest densities, the outcome of these experiments is qualitatively similar to the prediction of version 3, but this does not imply that low-ranked individuals are completely deprived of food, as assumed in version 3.

Experiments by Wilbur (1976) seem to confirm another interesting phenomenon: in spite of the claim by Wynne-Edwards (1962) and others that natural populations can become extinct due to overcrowding followed by the depletion of resources, we do not know of even one well-documented case of such an extinction in the field. We know about extinctions caused by adverse physical conditions, like poor weather or floods, or by the action of predators and diseases. Food may disappear because of various other reasons, and this can be followed by population extinction. We know of some cases of a population decline caused by food depletion in the field, but not of a complete extinction, although exhaustion of food followed by a complete extinction is common in laboratory experiments. The lack of well-documented descriptions of such extinctions in the field suggests that even if they do occur, they are extremely rare, and therefore versions 1 and 2, which predict such extinctions, only very rarely apply to field data.

The predictions given by version 4 can be illustrated by Donald's (1951) data relating the yield of clover *Trifolium subterraneum* to the density of its seeds sown (Figure 2.9). The yield increased with the number of seeds up to a seed density of 0.15 per square centimeter; above this level, and up to 3.2 seeds per square centimeter, the yield remained constant at about 0.9 g of dry matter. The number of seeds in the next generation may not be linearly proportional to the entire yield, because with increasing density the reproductive parts of the plants are the first to be affected (Harper 1977). Nevertheless, above a certain level, the density of populations of both plants and sedentary animals seems to be independent of population density in the previous generation. The data on clover presented here are a good illustration of this phenomenon, which has been called the "law of constant final yield" (Kira et al. 1953).

As mentioned earlier, there is a basic difference between cultivated plants such as maize, which was used to illustrate version 1, and wild plants, which are able to divide resources unequally. The prediction given by version 4, that above a certain density current density is independent of the density of a generation ago, seems to be so obvious to botanists that they tend to ignore population models developed by animal ecologists, who almost always assume the dependence of $N(t+1)$ on $N(t)$. The most interesting issue for botanists is the relation of the standing crop

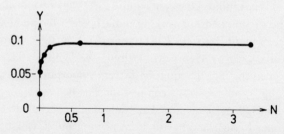

FIGURE 2.9. Total yield Y of *Trifolium subterraneum* in grams of dry weight as the function of the number N of seeds sown per square centimeter. Data from Donald (1951).

or production to environmental conditions, and not to population density of a generation ago. This approach seems reasonable, because it is difficult to imagine that a dense cover of wild vegetation or its seed production could be reduced by sowing more seeds.

The model of survival proposed in version 4 also can be applied to animals. Le Cren (1973), studying the survival of trout fry over the first five months of their life at eight different initial densities from several to 300 individuals per square meters, has found that at initial densities above ten individuals, the final densities are fairly constant, amounting to about eight surviving individuals per square meter. These results may be due to territorial aggressive behavior of the young fish. On the other hand, a similar survival pattern is also known to occur among animals that do not exhibit territorial behavior. For example, the number of surviving beetles *Stegobium panaceum* is independent of the initial egg densities ranging from 64 to 512 per gram of medium (Bellows 1981).

2.5. SCRAMBLE AND CONTEST COMPETITION

Since intrapopulation competition is almost always modelled on the logistic equation, contemporary ecologists pay relatively less attention to the mechanisms of competition within a single population than to interspecific competition. Fortunately, there are some exceptions. One can acquire a very good view of how intraspecific competition works from the book on the population ecology of plants by John L. Harper (1977), as well as from a series of papers by Henry M. Wilbur (see Wilbur 1980). In addition to the logistic equation, a very important concept of "scramble" and "contest" competition has been proposed by Nicholson (1954). I believe that Nicholson's concept deserves much more attention than it is receiving, and I will try to examine it now more closely, in order to show how it relates to the four versions of the model presented in section 2.2.

Nicholson (1954, pp. 19, 20) defines scramble and contest competitions in the following way:

> *Scramble* is the kind of competition exhibited by a crowd of boys striving to secure broadcast sweets, and is illustrated by the competition of *L. cuprina* larvae already described. Its characteristic is that success is commonly incomplete, so that some, and at times all, of the requisite secured by the competing animals takes no part in sustaining the population, being dissipated by individuals which obtain insufficient for survival. With *contest*, on the other hand, the individuals may be said to compete for prizes (such as a host individual, or an amount of favourable space an individual can arrogate to itself) which each provides as much of the requisite as an individual needs to enable it to reach maturity, or provides fully for the development of one or more of its offspring. Thus the individuals are either fully successful, or unsuccessful; and the whole amount of the requisite obtained collectively by the animals is used effectively and without wastage in maintaining the population.

A more formal description of these two kinds of competition was given by Varley et al. (1973), who defined them by comparing the number of individuals after competition as a function of their number before competition, without considering reproduction explicitly. Since competition influences both reproduction and survival, it seems appropriate to compare not the number before and after competition within the same generation, but rather the number in an earlier generation $N(t)$ and in a later one $N(t+1)$, as in the model presented here.

As far as contest competition is concerned, the model by Varley et al. does not differ from version 4 of the model presented here (Figures 2.3 and 2.4). If there is a sufficient amount of resources for an equilibrium number N_e of individuals, then only N_e individuals will survive and reproduce; all others will die without affecting the reproductive rate of the survivors.

The description of scramble competition by Varley et al. (1973) assumes that up to a certain density, say N_0, there is a linear relation between $N(t)$ and $N(t+1)$ (exactly like in all the versions of the model presented here), but if $N(t) > N_0$, then there is a crash and a complete extinction of the entire population. This is exactly the same outcome as given by function $k(N)$ for version 1 (Figure 2.3 [1]). It does not occur in version 2, in which scramble competition makes only k a decreasing function of N. Similarly, in both versions 1 and 2 scramble competition makes $N(t+1)$ a decreasing function of $N(t)$.

The terms "scramble" or "contest" refer to competition, i.e., to the situation in which there is a shortage of resources. In the model presented here, competition occurs when $N(t) > V/a$, so that according to the model presented here, the population size can either decrease or remain stable. Competition can also occur when the population is still growing, but at present I will ignore such a situation. It is possible to define scramble competition as one in which there is equal resource partitioning, but this is contradictory to the original definition by Nicholson (1954), since the blowfly larvae, presented as an example of scramble competition, do not divide their resources equally. What seems important in Nicholson's concept is the use of the resources by individuals that do not get enough to reproduce, which causes a decrease of population size. Scramble competition is therefore a process that, above a certain population size, makes population size $N(t+1)$ a decreasing function of the size $N(t)$ a generation ago. This does not imply that any decrease in population size or density is necessarily a result of scramble competition. What this definition requires is a decrease of population size resulting from the shortage of resources, which in turn is caused by high population density in relation to available resources. This seems to comply with Nicholson's original definition. In the model presented here, scramble competition occurs when $N(t) > V/a$ in versions 1 and 2, and when $V/a < N(t) < 2V/a$ in version 3 (Figure 2.4). In contest competition, some individuals are able to protect their resource intake, and therefore an increase in population density does not affect them or their reproductive abilities, so that $\mathcal{N}(t+1)$ is independent of $\mathcal{N}(t)$. This is the case for $\mathcal{N}(t) > V/a$ in version 4 and for $\mathcal{N}(t) > 2V/a$ in version 3.

These two kinds of competition can be defined more precisely in terms of individual resource intake y. In the case of scramble competition, an individual's resource intake y is not only the function of its position x within the population and of the amount V of resources available for the entire population, but is also determined by the population size $\mathcal{N}$. More precisely, individual resource intake $y(x, V, \mathcal{N})$ is also dependent on the number of individuals of lower ranks. In pure contest competition, this intake $y(x, V)$ is independent of $\mathcal{N}$: it can be low when the position x of an individual is low, but an increase in the number of individuals of even lower ranks does not affect the resource intake of those of higher ranks.

A more detailed discussion on the concept of scramble and contest competition is given in Chapter 6.

Individual Variation of Body Weight in Plant and Animal Populations

Body weight seems to be the most easily obtained hard data on which the estimation of individual resource intake and its variability can be based. Members of a plant or animal population often differ considerably in their body weights, even if they belong to the same sex and are of the same age. Differences are particularly pronounced among species exhibiting indeterminate body growth, like plants or fish, which continue to grow after attaining maturity. Nevertheless, organisms with determinate body growth can also be quite variable in their sizes. Three questions concerning these data seem to be of importance: (1) What does the individual variation of plants and animals look like? (2) What are the mechanisms causing this variation? (3) What are the consequences of this variation for the models of resource partitioning discussed in Chapter 2? An extensive review of individual variation in body weight of plants and animals has recently been made by Uchmanski (1985), and what is presented below is in part based on this review.

3.1. EMPIRICAL DATA

Uchmanski (1985) lists more than one hundred different plant and animal populations for which data on body weight distribution are available. These data differ in character; some are based on well-designed experiments that show how weight distribution changes with time and density in groups of organisms of the same age, while others are based on samples taken natural populations, about which very little is known. There is a large amount of data on the size distribution of trees in both natural and planted forests. Interesting data are the results of some studies made in Japan in the 1950s on plant intraspecific competition (Kira et al. 1953; Koyama and Kira 1956), data on plant growth reviewed by Harper (1977), and the results of experiments on the growth of amphibian larvae (Wilbur and Collins 1973; Wilbur 1976, 1980).

Following Uchmanski (1985), the available data can be characterized as follows. (1) Weight distributions are usually positively skewed (for example, see Figure 2.1), with small individuals being much more numerous than large ones, so that the median is smaller than the arithmetic mean. (2) Symmetric distributions, close to normal ones, usually occur at an early stage of growth. (3) These distributions become positively skewed when mean body size increases, but sometimes they remain symmetrical, and only very rarely do they become negatively skewed. (4) Higher density, food shortage, poor food quality, and adverse environmental conditions (e.g., pollution) accelerate the process of skewness increase and make the distribution more skewed up to the point when it becomes L-shaped. (5) Increasing mortality often accompanies high positive skewness. (6) There are more data on weight skewness than on its variability, but those available show an increase in both variance and coefficient of variation with both body size increase and increase of density in relation to available resources. Poor resource quality may also bring about an increase in the variance of body size.

Besides the data described above, there is another well-known biological phenomenon worth mentioning here: the quantal response of a group of organisms to the concentration of a drug or other unfavorable factor to which they are exposed. This is usually studied by counting the number of individuals in the group that have either died or changed qualitatively due to a quantitative change in their environment. These quantal responses vary considerably, so much so that they require special statistical methods, such as probit analysis (Hewlett and Plackett 1979). Not only does this method show that plants, animals, and microorganisms are variable in their susceptibility to adverse conditions, but it also suggests skewed distributions of this susceptibility, so that the logarithmic transformation of data is a standard procedure of probit analysis. This method seems to confirm what we know from the studies of individual weights: the variability increases and the distribution becomes more skewed when living conditions deteriorate.

The data mentioned above do show that there are differences in body weight among individual plants and animals that justify the assumptions about unequal resource partitioning used in Chapter 2. But the problem of individual variability is too important to be abandoned without any comment at this stage. Differentiation in weight and resource intake are phenomena that reflect the detailed mechanism of competition and that require something more than a strictly phenomenological description. The level of variability of body size differs among populations; with increasing body size, variation usually increases but occasionally decreases; weight distributions are usually skewed, but they may sometimes be symmetrical. It is known that some populations are stable while others oscillate violently; there may be other reasons for such oscillations, but if they are caused by the mechanisms described in Chapter 2, then it is important to investigate more closely under what circumstances one can expect high intrapopulation variability in body size, and what causes its low variability.

3.2. SOME SIMPLE EXPLANATIONS AND THEIR SHORTCOMINGS

The growth of an individual can be described by various mathematical models; however, at the early stage of growth, it can often be adequately described by the exponential equation in the form

$$W(t) = W(0)e^{rt}, \tag{3.1}$$

where $W(t)$ denotes the body weight at time t, $W(0)$ denotes body weight at the start of growth, and r is the constant coefficient, determining the rate of growth.

Koyama and Kira (1956) used equation (3.1) to explain the occurrence of the skewed distributions and the increase of variability of individual weights. They assumed that either $W(0)$, r, or both are normally distributed random variables. If $W(0)$ has a normal distribution and r is constant, then $W(t)$ is also a normally distributed random variable, with its standard deviation s increasing exponentially and proportionally to the mean body weight W_m. This implies that the coefficient of variation s/W_m does not change during growth. Thus it can be said that in this model, variation is not generated by the process of growth, since the variance increases proportionally to the mean.

If $W(0)$ is identical for all individuals, and r is a normally distributed random variable, then $\ln W(t)$ is also normally distributed, because from equation (3.1),

$$\ln W(t) = \ln W(0) + rt. \tag{3.2}$$

It follows from (3.2) that $W(t)$ is positively skewed, with a logarithmic normal distribution. In this case, the increase in variance is not linearly proportional to the average body weight, but is much higher, which means that the coefficient of variation increases with body size. This model allows for the generation of individual variation, but it requires differences among individuals other than body size alone: more precisely, it requires differences among individuals in the rate of growth r.

The above theoretical result accords with the theory of proportionate effect, describing stochastic growth, proposed by J. C. Kepteyn in 1903 (Aitchison and Brown 1957). Under exponential growth, the increase in body weight in a unit of time is proportional to the initial weight. This can be expressed by the equation

$$\Delta W = W(t+1) - W(t) = rW(t). \qquad (3.3)$$

According to this theory, if r is a random variable independent of $W(t)$, then the body weights $W(0)$, which are initially identical, will increase, their variability will increase, and their distribution will become more skewed.

The two models presented above explain the results obtained by Turner and Rabinowitz (1983) for the experimental monocultures of prairie grass *Festuca paradoxa*. Isolated plants were shown to exhibit a higher increase in skewness than crowded plants do. Competition among individuals of this species does not lead to higher variance and increased skewness as it does among many other plants and animals, although other effects of competition, such as smaller size, can easily be seen. Since the growth of the isolated plants was less inhibited than that of crowed ones, the increase in skewness may be higher among the isolated individuals than among crowded ones. Similar results are known for other plant species (Turner and Rabinowitz 1983), but we still know of many species of plants and animals in which higher skewness and variance result from competition between individuals.

Wilbur and Collins (1973) applied the following equation to describe the growth of some amphibian larvae:

$$W(t) = W(0) \exp\{(A/a[1 - \exp(-at) + u]\}, \qquad (3.4)$$

where A is the rate of exponential growth at time $t = 0$, a is the negative exponential rate of damping of this growth, and u is a normally distributed random variable with a mean equal to zero. This model of growth appears very complicated, but the increasing skewness and variation are simple consequences of u being placed in the exponent: $W(t)$ is logarithmic normally distributed, as in Koyama and Kira's (1956) model, and according to equations (3.1) and (3.2), we can expect an increase of both variation and skewness of distribution.

All three models have one feature in common: they explain both the increased variation and increased skewness of the distributions, but they do not explain how density or, more precisely, how the resource shortage due to increased density can produce an increase of variation and skewness. They do not describe competing individuals, and their application should be limited to groups of isolated individuals that do not interact with each other in any way.

If we ignore the skewness of the distributions of individual weights and apply a much simpler model of binomial or normal distribution, other aspects of individual variability will become clear. I will present here a simple stochastic model similar to that proposed by de Jong (1976) for the description of intra-population competition. Let us assume that in a discrete time unit each individual either obtains one unit of food with probability p or does not obtain it with probability $1 - p$, and that every unit of food gives a unit increase in body weight. With $W(0)$ identical for all individuals and, for simplicity, $W(0) = 0$, after T time units, $W(T)$ is binomially distributed or, for a large value of T, normally distributed with the mean

$$W_m = pT, \qquad (3.5)$$

and variance

$$s^2 = p(1-p)T. \qquad (3.6)$$

If food is superabundant, then $p = 1$, $s^2 = 0$, and at time T all individuals have the same weight $W(T) = T$. If food is scarce and p is smaller than unity, the variation in $W(T)$ increases. From equations (3.5) and (3.6), the coefficient of variation s/W_m is given by

$$s/W_m = \sqrt{(1-p)/(pT)}, \qquad (3.7)$$

which implies that variation is a decreasing function of p.

This model explains the increase of variation with a decreasing amount of resources. Thus it can be said that competition for resources results in higher individual variation; but the model explains neither the skewed distribution nor the increase of variance with time. As a matter of fact, equations (3.6) and (3.7) clearly show that with increasing time T, both the variance and the coefficient of variation decrease.

The phenomenon of unequal resource partitioning, especially among plant populations, is also known as size hierarchy. Weiner and Solbrig (1984) have recently argued against the application of positive skewness as a measure of inequality among plant individuals, and they have proposed measures of inequality that are applied in economics, the so-called Lorenz Curve and the Gini Coefficient. They are obviously right in claiming that skewness does not measure inequality and that these two economic measures reflect our intuitive understanding of inequality. I will not describe these measures here, but I would like to point out that they do not contain more information than is given by the distribution of resource intakes y or of individual weights. The Lorenz Curve can be applied instead of the function $y(x)$ to describe the relation of sizes or resource partitioning within a single-species population, but this curve cannot be used to calculate easily how many individuals are above or below a certain critical size or resource intake. Weiner and Solbrig are also correct that positive skewness does not always result from competitive interactions among individuals within a population. On the other hand, skewness is an important feature of the distribution of sizes of both plants and animals that cannot be ignored and that deserves an explanation.

To make the process of growth differentiation under intra-population competition understandable, a model is required that will explain the increase of both variation and skewness of distributions, with growth and with the gradual depletion of available resources. Such a model should also consider the phenomena that have been ignored here, namely, plants and animals do not use resources proportionally to their availability, and they do not grow to infinity. There is a satiation level of resource intake and a limit to the growth of body weight. These two phenomena are expected to act toward uniformity, that is, to decrease the variance between individuals. We need a model to explain not only how individual variability is generated, but also under what circumstances one can expect plants and animals to become more uniform.

3.3. WEIGHT DIFFERENTIATION UNDER STOCHASTIC GROWTH

There are mathematical difficulties in constructing a model that would explain skewness and variation. These difficulties can easily be overcome through numerical simulations, but such simulations do not provide the real general explanation; they can only serve as an additional aid to understanding the problem better.

An example of a numerical simulation is presented in Figures 3.1 and 3.2. At the beginning, one hundred individuals, each of the weight $W(0) = 1$, are supplied with $V(0) = 1000$ food particles. In a discrete time unit, an individual can either catch one food particle and increase by one unit of weight or not catch it and remain at the same weight. The probability $P(t, i)$ that in the tth time unit an individual i will catch a particle is given by

$$P(t, i) = 1 - \exp[-cV(t)W(t, i)], \qquad (3.8)$$

where c denotes a constant coefficient that equals 10^{-4} for the simulation presented here. In every time unit t the amount $V(t)$ of resources is decreased by the amount taken by all individuals. The probability of catching a food particle given by equation (3.8), is an increasing function of the amount of food available $V(t)$ and of the size of an individual $W(t, i)$, but it also takes

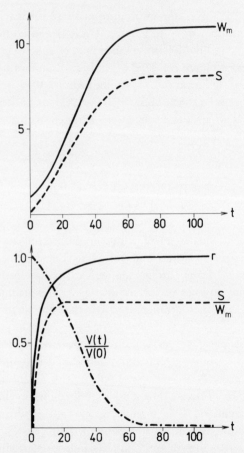

FIGURE 3.1. Results of the stochastic simulation of the growth of 100 individuals in 110 time units. The arithmetic mean of the body weight W_m, its standard deviation s, coefficient of variation s/W_m and correlation r, as well as proportion of resources left $V(t)/V(0)$, are presented as functions of time t. From Łomnicki (1980b).

into account the satiation phenomenon, since no more than one food particle can be taken by an individual during a time interval. This makes the simulation presented here different from the theory of proportionate effect (section 3.2), where no satiation is assumed. This also brings the function $P(t, i)$ as defined

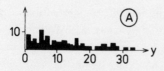

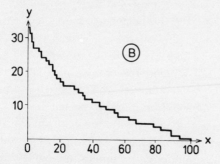

FIGURE 3.2. The final result of the simulation presented in Figure 3.1, after 110 time units. The variability of the final sizes $W(110)$ of individuals, which are linearly related to the amount y of resources taken by each individual, is presented here as (A) frequency distribution, and (B) the function $y(x)$. From Łomnicki (1980b).

by equation (3.8) close to the functional response of predators as defined by Holling (1959).

It has taken 110 units of time for these 100 individuals to use all 1000 food particles. The changes in the mean weight W_m, its standard deviation s, the coefficient of variation s/W_m, and the proportion of food used $V(t)/V(0)$, are presented in Figure 3.1. To find out to what extent the fate of an individual is determined in early stages of its growth, the correlation coefficient r between the body weights of individuals at each unit of time, $W(t, i)$, and their final body weights, $W(110, i)$, was calculated. Note that the correlation coefficient r grows very rapidly (Figure 3.1) during the first twenty time units, yielding 0.9, which means that the total amount of food that an individual takes during its life is determined early, when there is still plenty of food for the entire population. The distribution

55

of body weights at the end of the simulation is presented in Figure 3.2 in two different ways: as a frequency distribution (A), and as the function $y(x)$ (B) as described in section 2.1. These two diagrams show once again the relationship between frequency distribution and the function $y(x)$. The arithmetic mean of the weights at the end of the simulation was 11 (10 food particles taken by the average individual + initial body weight $W(0) = 1$), and the median weight was 9, which means that the distribution was positively skewed.

To make this simulation more realistic, one should include maintenance cost and limited growth, as well as a possibility of continuous inflow of food into the system. However, it is much more important to construct a general analytical model of weight differentiation. Such a model was proposed by Kimmel (1986). His assumptions were similar to those in the simulation presented above: an individual can catch no more than one food particle in a unit of time (satiation effect), and the probability of catching it depends on the individual's weight. In the simple version of his model, one food particle is offered to a group of individuals in each time unit; if one individual takes it, the others are left without food in this time unit. This results in increased variation and skewed distribution of body weights. The same is true in a more generalized version of the model, in which more food particles are offered, but still fewer than the number of individuals in the group. Kimmel showed that skewed distribution cannot be achieved unless intraspecific competition is included. In other words, the presence of individuals that affect the food intake of others is a necessary condition for the skewed distribution to occur. His analytical model is simple, but its conclusions are general and very important: if there is a saturation effect, proportional growth does not make the distribution skewed; competition among individuals is required to bring about skewness and increasing variation.

An empirical counterpart of the stochastic models discussed here can be found in the example of variation in growth within natural populations of the intertidal barnacle *Balanus balanoides*

(Wethey 1983), in which the growth of an individual is determined by the number of food particles it acquires. These data seem to confirm the predictions of the model, because the distribution of body size is skewed and there is a correlation between body sizes of the same individuals in June and December. The skewness is also higher for individuals that have more neighbors within one square centimeter. On the other hand, skewness does not increase with time and with body growth, and the correlation mentioned above is rather low, so that the size of the opercular area of an individual in June in a rather poor predictor of its size in December.

3.4. DETERMINISTIC GROWTH AND THE IMPORTANCE OF EARLY DIFFERENTIATION

Uchmanski (1985) has proposed a model of weight differentiation in the process of deterministic growth that applies the following growth equation, based on the energy budget

$$dW/dt = af(V, W_0)W^b - cW, \qquad (3.9)$$

where a, b and c are constant coefficients of growth $(0 < b < 1)$ and the function $f(V, W_0)$ describes the effects of resource abundance V and of the initial body weight W_0 on growth. If one starts with different initial weights, it is impossible to arrive at increased variation and a skewed distribution without the additional effect of $f(V, W_0)$. Increasing variation and skewed distribution related to the shortage of resources would be possible if one assumed that individuals of higher initial weight had some advantage in addition to their weight, for example if they had access to better or more abundant resources. One difference between Uchmanski's model, which makes differentiation and skewness possible, and the earlier deterministic models by Koyama and Kira (1956) is that Uchmanski's model is able to generate skewness and variation by resource shortage. Another important difference is that Uchmanski's model considers limited growth, under which an increasing differentiation of body size is less likely.

It is an open question whether the model of body growth based on the energy budget, as given by equation (3.9), describes the growth of the body alone or both body growth and progeny production. According to the optimization model by Ziolko and Kozlowski (1983) of how resources should be divided between growth and reproduction, growth should be arrested much below the size at which assimilation is balanced by respiration, so that large amounts of energy can be used for reproduction. If this is so, then a different model is required to describe growth alone. Nevertheless, since the basic properties of the process of growth are preserved in equation (3.9), the conclusions that Uchmanski (1985) reached seem generally valid.

Even without any mathematical model it seems obvious that deterministic weight differentiation is impossible without individuals being different at the beginning of growth or becoming different at a certain point of growth. There are data that support the importance of initial body weights for future growth. The difference in germination time of seeds of *Dactylis glomerata* can, according to Ross and Harper (1972), account for 95 percent of the variance of individual weight at harvest time, and there is quite a close relation between emergence ranking and mean weights at harvest time of those individuals that germinated on the same day. Similar data are available for animals: nestlings of birds of prey exhibit large differences within a single brood (Ingram 1959), with those hatched earlier taking a disproportionally larger share of food and having a much higher chance of surviving. This phenomenon is not limited to birds of prey; it has been reported for swallows (Bryant 1978), some mammals, and insect larvae, for example.

These differences present at a very early stage of life may be environmental, with some individuals incidentally taking better places than others, but there are two other reasons for the existence of individual differences at the early stage of plant

58

and animal life: (1) the ability of plants and animals to produce variable progeny, and (2) hereditary differences in germination time or in the rate of early development in a given environment. The selective advantage of the existence of such differences is discussed in section 4.1.

3.5. WEIGHT DIFFERENTIATION IN COMPETITION FOR SPACE

The models discussed above have not considered the relations between individuals competing with their neighbors for space or light, a phenomenon common in terrestrial plants. Numerical simulations of this process have been presented by Wyszomirski (1983), who considered individuals, each occupying a circular area, with the size of the individual linearly proportional to the area of such a circle. Weight increase $\Delta W(t)$ in a time unit t is determined in this model by the equation

$$\Delta W(t) = A(t) - R(t), \qquad (3.10)$$

where $A(t)$ and $R(t)$ denote assimilation and respiration, respectively. Whereas respiration is linearly proportional to the individual's size $W(t)$ and is not influenced by neighboring plants, assimilation includes this influence in the following way: with increasing body size, there is a linearly proportional increase of the circular area occupied by the individual, which results in overlapping of the surfaces occupied by neighboring individuals, and consequently the assimilation $A(t)$ is n times smaller for those parts of the circle surface on which n individual circles overlap.

Applying these simple rules, the growth of 100 individuals during 80 units of time was simulated, for several different densities and for both random and clumped dispersion patterns. All simulations started with a cohort of individuals of identical initial size $W(0)$. If the dispersion pattern of individuals is uniform, no growth differentiation is possible, since overlapping

59

between neighbors is identical for each individual. The same is true when density is low because individual areas do not overlap. With a random dispersion pattern, individual sizes are, at the beginning, negatively skewed; the coefficients of variation and of skewness increase with time and with increasing density (Figure 3.3). With clumped dispersion, the distribution of sizes is already positively skewed after 20 units of time and, as in the case of random dispersion, both the coefficient of variation and the coefficient of skewness increase, although, as shown in Figure 3.3, they are higher for clumped than for random dispersion.

These simple simulations show that the variation in body size and the skewness of its distribution can be attributed to competition, expressed by the overlapping of areas controlled by individuals, provided these individuals have a nonuniform dispersion pattern. In other words, individual variation and skewness are functions of population density and nonuniform distribution.

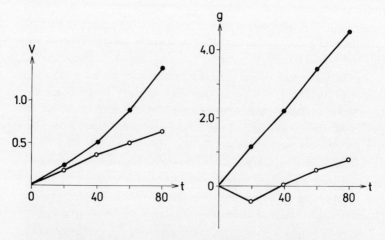

FIGURE 3.3. Coefficients of variation V and of skewness g during consecutive time moments t, as determined by the simulation of competition among individuals that represent either random (open circles) or clumped (black circles) distributions. Data from Wyszomirski (1983).

3.6. WEIGHT DISTRIBUTION AND
GENERAL PROPERTIES OF THE FUNCTION $y(x)$

Although there is a nonlinear relationship between body weight and individual energy intake, we may suppose that the general shape of the distribution of individual energy intake y is similar to weight distribution, except that energy intake should be more variable for two reasons: (1) the less efficient conversion of energy into body weight for individuals with the highest intake, and (2) the absence from the data of the weights of these individuals that have died earlier because of an insufficient energy intake.

In the model presented in Chapter 2, the maintenance cost m was identical for all individuals, and the difference $(y - m)$ was linearly proportional to individual reproductive output. The real maintenance cost is weight-dependent; therefore m should decrease with decreasing y. What is of importance here is not the maintenance cost itself but the minimum energy requirement for the survival of an individual. Such a requirement should vary less between individuals than their consumption does. We can also expect an individual to die not because its assimilation is below its respiration for the entire lifespan, but because respiration exceeds assimilation plus energy reserves at a certain moment of time. To develop a realistic model of individual growth within the population, one may follow the simulation model developed by Fujii (1975) for a cohort of insects. Unfortunately, he provides no information as to how the individual differences may affect the fate of the cohort.

Nevertheless, even without a detailed description of each individual, it is possible to see how the versions of resource partitioning presented in section 2.2 can be modified to account for the data on weight distributions. At low population density, when resources are superabundant, so that $V > a\mathcal{N}(t)$, variation of resources intake y is not important for the population dynamics, as long as the reproductive output is linearly proportional to $(y - m)$. This is because if $y > m$ for each individual, then $\mathcal{N}(t+1)$ as defined by equation (2.3) is the same, irrespective of whether $y = a$ is identical for each individual or whether it is the arithmetic mean of a random variable. If there is a shortage of resources, individual variation is of fundamental importance for population dynamics, and we can expect individual food intake y to be a variable with skewed distribution (Figures 2.1 and 3.2). After comparing the function $y(x)$ as presented in these figures, with $y(x)$ given by a straight line in versions 2 and 3 (Figure 2.2), some conclusions can be made for the skewed distribution, as presented below.

The resource intakes of individuals with highest ranks (lowest x values) are much higher than those given by the straight line (Figure 2.2 [2]), which implies that extinction of the entire population due to resource shortage is less likely to occur in reality than in version 2. The data on body weights do not confirm the assumption used in version 3, that the individual with the highest rank should obtain the maximum possible resource share a (Figure 2.2 [3]); nevertheless, a decrease in body weights of those of highest ranks due to increasing density is relatively small, so that this simplified assumption is not far removed from reality. From the weight distributions one can see that a resource shortage affects the majority of population members, but primarily those of lower ranks, not those of higher ranks.

Positively skewed distributions result in a much larger proportion of individuals with a very low resource intake y than that given by the straight line function $y(x)$. This implies that there are more individuals with $y < m$, using resources for maintenance only, not for reproduction. Generally, the skewed distribution should produce a pattern of population dynamics that lies somewhere between version 2 and version 3, i.e. with higher population persistence than in version 2 and with lower values of $\mathcal{N}(t+1)$ for $\mathcal{N}(t) > 2V/a$ than in version 3.

From the data on weight distribution and the preliminary remarks in section 2.1, one can deduce that y as a function of individual rank x, the amount V of resources for the entire population, and population size $\mathcal{N}$ should have the following properties. (1) by definition $y(x, V, \mathcal{N})$ is a nonincreasing function of x, since we have to include the possibility that y is identical for several population members, and (2) the sum of y's from $x = 1$ to $x = \mathcal{N}$ is smaller than or equal to V. An increase in the total amount V of resources results in an increase of the intake of each individual, thus (3) y is an increasing function of V and, more precisely, taking into account the saturation effect, it is a nondecreasing function of V. Because of this satiation effect, (4) dy/dV is a nonincreasing function of V with $y \leq a$. Properties (3) and (4) can also be due to other factors (such as the time of handling prey by a predator) that determine the functional response of predator, as defined by Holling (1959). Note that the functional response refers to the number of prey taken by the population of predators, while $y(x)$ determines the number of prey taken by an single individual of rank x. The variation in y is given by the slope of $y(x)$, and since this variation increases with food shortage, (5) $|dy/dx|$ is a decreasing function of V.

These five properties hold for both scramble and contest competition, but for scamble competition two additional properties must be added to account for the ability of low-ranked individuals to decrease the individual resource intake of all population members. These are: (6) y is a decreasing function of $\mathcal{N}$, and (7) $|dy/dx|$ is an increasing function of $\mathcal{N}$.

These seven general properties of the function $y(x, V, \mathcal{N})$ can be applied for the prediction of population dynamics (Uchmanski 1983; Łomnicki and Ombach 1984), but the predictions are very general. At the present state of our knowledge of resource partitioning, it is rather difficult to present a realistic formula for this function. It cannot be derived from the logarithmic normal distribution, since the theory of proportionate effect on which this distribution is based ignores the saturation effect and the limitation of growth. Even if we accept that y is lognormally distributed, no explicit formula for $y(x, V, \mathcal{N})$ can be derived. Such formulae have been proposed for contest competition when $y(x, V)$ is independent of $\mathcal{N}$. The first attempt

(Łomnicki 1978) was to apply a geometric series in the form

$$y(x, V) = a(1 - a/V)^x. \qquad (3.11)$$

This formula implies that each individual takes a units of food, diminished by the fraction a/V by every individual of higher rank. It can be applied to a group of predators that ambush along a path on which prey moves in one direction, so that individuals positioned further along this path take what has been left by those of higher ranks which are closer to the entrance of the path.

Gurney and Nisbet (1979) have proposed a more flexible formula:

$$y(x, V) = AV/(1 + AV/a + gx), \qquad (3.12)$$

where A and g are parameters allowing for more or less unequal resource partitioning. In formula (3.11), variation in resource intake y was determined solely by the resource abundance V in relation to the maximum possible intake a.

These two formulae have a significant shortcoming: they are not general enough. Rank x means here a rank in a linear hierarchy, and it cannot be applied to a group of individuals that differ only in their resource intake. The function $y(x)$ is more than just another way of presenting the distribution of resource intakes y. This is because $y(x, V)$ is not equal to $y(2x, 2V)$. To apply this formula, we have to identify a group in which all individuals are related by rank, and we cannot apply it to a random sample of such a group or a population made of several such groups. Besides, these formulae allow for contest competition only, because y is independent of N.

As discussed in section 2.5, the stability of populations depends on whether competition is scramble, with y determined by N, or contest, with y independent of N. This can hardly be determined using body weight distributions and will be discussed more extensively in Chapter 6.

CHAPTER FOUR

Individual Differences and Hereditary Variation

So far, I have discussed differences that can arise in a group of genetically identical individuals, of the same age and sex. Such differences are usually ignored by theoretical ecologists. However, ecologists have written quite a lot about age-structured populations and about the age dependence of natality and mortality. The genetic variation of populations has also recently been of interest to them. This chapter does not review the genetic aspects of ecology; it is intended only to discuss heredity as a mechanism generating individual variation, which is important for population dynamics. Individual differences caused by age and models of population dynamics with overlapping generations are discussed in Chapter 5.

The progress made in population genetics over the last twenty years has prompted us to reject the classic hypothesis of the uniformity of natural populations. We are now inclined to adopt the balance hypothesis of great genetic variability, maintained by various kinds of frequency-dependent selection (Lewontin 1974). The progress in evolutionary ecology resulting from the introduction of the concept of the evolutionarily stable strategy (Maynard Smith 1982) has been tending toward the same direction, denying that genetically uniform individuals living in an identical environment must necessarily exhibit the same behavior. The possibility of the existence of mixed strategies means that not only the mean value of a trait but also its variation may be adaptive. Some applications of game theory to a few simple models of population self-regulation and to emigratory behavior are given in sections 7.1 and 8.1, respectively.

CHAPTER FOUR

The application of decision theory (section 4.1) shows even more clearly that variation among individuals can be an important adaptation, while the concept of soft selection (section 4.2) enables us to relate genetic selection to the models of population dynamics presented in Chapter 2. Other aspects of the genetic determination of intrapopulation variability are given in section 4.3.

4.1. VARIATION AS AN ADAPTATION

In Chapter 3, the variation in body size of plants and animals was seen as a by-product of their growth on limited resources. To generate this variation it is necessary to assume random food intake, random spacing, or habitat heterogeneity. The initial variation of newborn individuals, seeds, or eggs may be another cause of variation in body size. This initial variation may be inherited, or it may arise during the development of the zygote in the parental organism, but regardless of its proximate causes, such variation may also be of adaptive value.

The importance of individual variation for population stability was recognized by Wallace (1977, 1982). In his earlier paper, Wallace stressed the importance of individual variation for population persistence. Later (1982), he argued that since phenotypic variation is sometimes higher among the progeny of the same parents than among unrelated individuals, it should be considered an adaptation, because in the case of a shortage of resources, it may secure the survival of at least a part of the progeny.

The adaptive significance of variation can also be explained by the concepts of the fitness set (Levins 1968), the evolutionarily stable mixed strategy (Maynard Smith 1982), or the optimal decision theory under unpredictable conditions (Cooper and Kaplan 1982). This last concept is especially interesting because it explains the existence of initial differences that can later be amplified during the growth of individuals. Evolutionarily stable mixed strategies can exist if the success of an individual is frequency-dependent, i.e. if it depends on the strategies adopted by other individuals. The optimal decision concept does not require such an assumption.

The application of the optimal decision theory can be presented for an example of seeds that germinate early (E) or late (L) and that encounter a cold (C) or warm (W) spring (Table 4.1). Let us assume that a warm spring occurs at random with the probability $p = 0.6$, and therefore cold springs occur with the probability $(1 - p) = 0.4$. Assuming nonoverlapping generations, the reproductive success of these seeds can be described by their net reproductive rates R, which depend both on their

TABLE 4.1. A numerical example showing the selective advantage of a mixed M strategy over two pure strategies of early E and late L seed germination, during warm W and cold C springs

generation t	1	2	3	4	5
kind of spring	W	C	W	W	C
net reproductive rates for seeds germinating early E	2.0	0.1	2.0	2.0	0.1
the product of the net reproductive rates after t generations	2.0	0.2	0.4	0.8	0.08
net reproductive rates for seeds germinating late L	0.5	2.0	0.5	0.5	2.0
the product of the net reproductive rates after t generations	0.5	1.0	0.5	0.25	0.5
net reproductive rates for mixed M strategy: $q = 0.5$ seeds germinate earlier, while $(1 - q)$ germinate later	1.25	1.05	1.25	1.05	1.25
the product of the net reproductive rates after t generations	1.25	1.31	1.64	2.05	2.15

germination time and on the kind of spring they encounter. Let us assume a net reproductive rate $R_{WE} = 2.0$ after a warm spring, and $R_{CE} = 0.1$ after a cold one, for the seeds that germinate early. Applying the numerical example proposed here, one can easily see (Table 4.1) that for each seed germinating early, there will be, on average, $2.0^3 * 0.1^2 = 0.08$ seeds after 5 generations. The mean reproductive rate R_{mE} for seeds germinating early is $0.08^{1/5} = 2.0^{3/5} * 0.1^{2/5} = 0.60$. In order to predict the reproduction results after 5 years, the geometric mean of the reproductive rates during these 5 years is applied here, because the reproductive rates in consecutive years are not added but multiplied. The above numerical example clearly shows that after many generations, the mean net reproductive rate for the seeds germinating early is given by

$$R_{mE} = (R_{WE})^p (R_{CE})^{(1-p)}. \qquad (4.1)$$

Let us assume that seeds that germinate late are doing better in the cold season, so that, for example, $R_{WL} = 0.5$, while $R_{CL} = 2.0$. According to a formula analogous to equation (4.1), the mean reproductive rate for seeds germinating late, R_{mL}, would be equal to $0.5^{1/5} = 0.87$.

Instead of using one pure strategy of either early or late germination, a seed can apply a mixed (M) strategy, which can be defined as early germination with probability q and late germination with probability $(1 - q)$. According to Cooper and Kaplan (1982), the mean reproductive rate of such a mixed strategy is given by the equation

$$R_{mM} = [qR_{WE} + (1-q)R_{WL}]^p [qR_{CE} + (1-q)R_{CL}]^{(1-p)}. \qquad (4.2)$$

By finding the maximum of the function R_{mM} with respect to q, one determines the optimal probability q of early germination, which gives the highest possible mean net reproductive rate R_{mM}. For the numerical example presented here, the optimal q equals 0.50, which means that half of the seeds should germinate early, while the remaining half should germinate late. This gives a net reproductive rate $R_{WM} = 1.25$ if the spring is warm and $R_{CM} = 1.05$ if the spring is cold. When the probability p of a warm spring is 0.6 as assumed here, the mean reproductive rate for mixed strategists $R_{mM} = 2.15^{1/5} = 1.17$, which is much higher than such a rate for either one of the pure strategies.

Following Cooper (1981), the problem introduced here by the simple numerical example can be also presented in a graph, called a decision tree in the decision theory (Figure 4.1). With

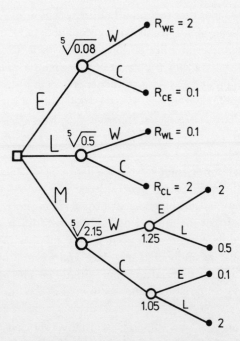

FIGURE 4.1. A decision tree for the hypothetical decision problem discussed in the text and presented in Table 4.1. The square node is the decision node, the round nodes are chance nodes, and the paths emerging from a chance node represent a set of mutually exclusive and exhaustive events whose probabilities sum to one. The rightmost black nodes are consequence nodes, at which the net reproductive rate for the particular strategy after a particular spring is given. When the mixed strategy is averaged over five different springs, the geometric mean is applied, but when it is averaged over one season, the arithmetic mean is used. These means are written at the chance nodes. The mean reproductive rate for a mixed strategy averaged over different kinds of spring is much higher than the respective rates for any one pure strategy.

different kinds of springs (warm, cold), the average reproductive rate is a geometric mean, because the springs occur in a succession and the rates have to be multiplied according to equation (4.1). This is quite different from the case of a mixed strategy, in which early germination is adopted with probability q, irrespective of the kind of spring a seed encounters. Kaplan and Cooper (1984) have called this "adaptive coin-flipping." "Coin-flipping" has the fraction q of seeds germinate early, while the fraction $(1 - q)$ germinates late. The mean reproductive rate of individuals adopting this mixed strategy within one season is an arithmetic mean, as shown by equation (4.2). Since the arithmetic mean of a set of non-negative variables is always higher than the geometric mean of the same set, the strategy of "coin-flipping" always gives better results in a variable and unpredictable environment than the application of any one pure strategy. This is especially apparent if at least one of the reproductive rates for a pure strategy in one kind of season is equal to zero.

Note that the reasoning presented here regarding the strategy of "coin-flipping" is very general, since it is based on the difference between arithmetic and geometric means of a set of variables. It can be applied not only to two different springs and two different strategies, but to any number of different hazards that an individual may face and for any number of different strategies adopted to cope with them. This allows also for continuous variation to be interpreted as a result of "coin-flipping" strategy.

The strategy of "coin-flipping" is intuitively obvious if applied to an individual that produces many offspring that face an unpredictable environment. This individual would do best to impose one pure strategy upon some of its offspring and a different pure strategy upon the remaining ones. This mixed strategy, however, does not seem intuitively the best one when a single individual chooses a strategy by "coin-flipping" instead of sticking to the one inherited from its parents. To explain more clearly the essence of "adaptive coin-flipping," I will present here two imaginary cases that I hope will make the point clearer and will also show the limitations of the method.

Let us consider an individual of a species with overlapping generations that lives in an unpredictable environment for many years, selecting for itself a strategy for each year and producing each year the number of offspring determined by the strategy chosen and the conditions encountered. Applying the symbols introduced earlier, the number of progeny produced by this individual during its lifetime is given by the equation

$$R_m = pqR_{WE} + p(1-q)R_{WL} + (1-p)qR_{CE}$$
$$+ (1-p)(1-q)R_{CL}. \qquad (4.3)$$

Since the different strategies and different kinds of season occur independently of one another, the probability of the occurrence of a given reproductive rate is the product of the probability with which a given kind of season occurs and the probability of adopting a given strategy. Here we are talking not about the mean reproductive rate for several generations but about the sum of all the offspring produced by one individual; therefore, the mean number of offspring produced during one season is the weighted arithmetic mean for all the seasons. Since R_m as defined by equation (4.3) is the arithmetic mean of pure strategies, it is not better than the best pure strategy. Thus, in this case, "coin-flipping" is not better than sticking to a pure strategy. This is because the value being maximized here is the mean reproductive rate not for several generations, but for one generation only, lasting for many different seasons. Therefore we do not apply the geometric mean, as is the case when we consider several generations.

Let us now consider the second imaginary case of a very strange "coin-flipping" strategy, in which in every generation all the population members select an identical strategy by "coin-flipping". This means that the strategy is mixed among generations but not among individuals. After many generations, the mean reproductive rate for such a mixed strategy is given

71

by

$$R_m = (R_{WE})^{pq}(R_{WL})^{p(1-q)}(R_{CE})^{(1-p)q}(R_{CL})^{(1-p)(1-q)}, \quad (4.4)$$

and undoubtedly such a mixed strategy does not give better results than the best pure strategy. This implies that such a mixed strategy cannot evolve by natural selection. This imaginary example clearly shows that the mixed strategy is not based on "coin-flipping" itself, but on the independent "coin-flipping" of every individual, resulting in the fraction q of individuals adopting the strategy of early germination, while the remaining fraction $(1-q)$ adopts the strategy of late germination.

As equation (4.3) shows, a single individual does not necessarily increase its lifetime reproductive success by "coin-flipping." Nevertheless, for many generations in an unpredictable environment, "coin-flipping" is the best replicating strategy. This confirms Dawkins's (1982) view that natural selection selects not for the best individuals, but for the best strategy. In the numerical example presented here, the net reproductive rate for the pure strategy of late germination never falls below 0.5, while for the mixed strategy of "coin-flipping" it falls to 0.1 in $(1-p)q = 0.2$ cases. In spite of this, the strategy of "coin-flipping" replicates better than the strategy of late germination.

There is obviously no discrepancy between the action of natural selection towards more efficient reproduction of individuals and the existence of large differences in reproductive success among these individuals. As shown by Kaplan and Cooper (1984), such large differences and low reproductive success of some individuals can be easily explained by "adaptive coin-flipping," provided the environment is variable in time and unpredictable.

4.2. DIFFERENTIAL MORTALITY AND THE SOFT SELECTION CONCEPT

Consider two genotypes A and B, such that under given environmental conditions genotype A survives better than genotype B. Say genotype A dies with the probability $m_A = 0.1$, genotype B with the probability $m_B = 0.3$. Assuming equal fertility of these two genotypes, we may use the probabilities m_A and m_B to estimate the relative fitness W of the genotypes in the following way:

$$W_A = 1,$$

$$s_A = 0,$$

$$W_B = (1 - m_B)/(1 - m_A), \text{ and}$$

$$s_B = 1 - W_B.$$

This is the basis of the formal description of selection due to differential mortality, but it does not contain information on the causes of this differential mortality or on the mechanisms behind the different proportions of surviving individuals. If conditions are deteriorating and there is a decrease in the survival due, for example, to increased density in relation to available resources, we do not know how the additional mortality, say m_D, affects different genotypes. The same is true for age-specific mortality: if there are individuals of two different age classes within the same population, we may determine their survival under given environmental conditions, but we cannot predict their survival when the conditions are improving or deteriorating.

We can imagine three different models (Figure 4.2) that describe the distribution of the additional mortality among two different groups within the same population. Each one yields different results. The first model assumes that an additional source of mortality m_D affects both groups in the same way, so that the new increased mortalities m_{AN} and m_{BN} for the genotypes A and B, respectively, are given by equations

$$m_{AN} = m_A + m_D - m_A m_D, \tag{4.5}$$

$$m_{BN} = m_B + m_D - m_B m_D. \tag{4.6}$$

These equations can be applied under the assumption that the probability m_A of death for an individual of genotype A under more favorable conditions is independent of the probability m_D of death due to additional sources of mortality when conditions are less favorable. For example, if death due to poor weather occurs with the probability m_A and the additional cause of death, such as predation, occurs with the probability m_D, then we may assume that these two causes act independently. Since these two causes of death do not exclude each other, the product of the probabilities must be subtracted from their sum. If the additional mortality is added according to equations (4.5) and (4.6), the relative fitnesses W and selection coefficients s do not change; hence $W_{AN} = 1$, while

$$W_{BN} = (1 - m_{BN})/(1 - m_{AN})$$
$$= (1 - m_B - m_D + m_B m_D)/(1 - m_A - m_D + m_A m_D)$$
$$= (1 - m_B)/(1 - m_A) = W_B.$$

This simple model of selection (Figure 4.2[1]) seems the most reasonable one if nothing is known about the population to which it is applied besides the coefficients m_A, m_B, and m_D. On the other hand, this model does not allow for the description of frequency-dependent selection or any other kind of selection without constant selection coefficients.

From the ecological point of view, the most interesting alternative to this model is the model of "soft selection" as proposed by Wallace (1968, 1975, 1981). The soft selection concept is an important one, because it allows the process of selection to be determined by population density in relation to available resources.

It is a well-known phenomenon that there are genotypes that under given environmental conditions are lethal, irrespective of how many resources their carriers obtain. It can be said that selection against the lethal genotypes does not depend on their frequency or on population density. Selection described by the model presented above is also both density- and frequency-independent, because additional mortality—which can be due, for example, to higher density—does not alter selection coefficients. Such selection is, according to Wallace (1975), classified as "hard selection." Hard selection can be easily administered

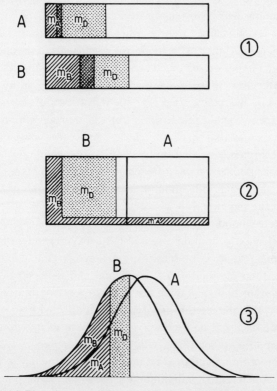

FIGURE 4.2 Three interpretations of selection against genotype B, which is assumed to represent lower fitness than genotype A. The symbols m_A and m_B denote mortality of the genotype A and B respectively, while m_D is an additional mortality due to an increased density in relation to available resources or to other adverse environmental factors. In the classic hard selection model (1), the additional mortality affects both genotypes equally. In the soft selection model (2), only the genotype B is affected by the additional mortality. The third interpretation (3) allows for the variation of the genotype's survival ability, and the additional mortality is determined by the distribution of individuals along the gradient of the survival ability.

among laboratory populations or among cultivated plants and domesticated animals by removing a constant fraction of a given genotype, but in nature both reproduction and survival are usually density- and frequency-dependent.

Soft selection can be described by an example of a population that consists of individuals of a superior genotype A and an inferior one B. Under extremely favorable conditions, at low population density in relation to available resources, we may expect no mortality to occur other than that due to random accidents or senility. As a result, the coefficients of selection for both genotypes are equal to zero. With a population increase and a decline in the amount of resources available for each individual, we may expect that the individuals of the inferior genotype will be the first to suffer from the shortage of resources. If one assumes that there are clear-cut differences between these two genotypes, such that under deteriorating conditions due to increased population density each individual of genotype A does better than each individual of genotype B, then the mortality affects the individuals of genotype B first, and only after their complete elimination does it start to affect those of genotype A. Soft selection is both density-and frequency-dependent for the following reasons: if the density in relation to available resources is low, then the mortality is accidental and the coefficients of selection for both genotypes are equal to zero. With increasing density, it is the inferior genotype that is affected first, so that if its frequency is high only a small fraction of individuals of this genotype is removed; with decreasing frequency a larger fraction of this genotype is removed; and if frequency is low, it may happen that all individuals are removed. At this point, the coefficient of selection for the inferior genotype is equal to unity.

How can one interpret the mortality of genotypes A and B, discussed above, in terms of soft selection? Since both genotypes are affected, we may assume the mortality m_A of the superior genotype A to be accidental, affecting both genotypes equally, while selection acts only against the weaker genotype B, removing the proportion $(m_B - m_A)/(1 - m_A)$ of individuals (Figure 4.2 [2]). Consequently, an additional mortality m_D should affect the inferior B genotype only, up to the point of its total elimination. The predictions concerning the distribution of additional mortality in the model of soft selection are quite different from those in the first model. Both models represent extreme cases: in the first, mortality is determined by the genotype alone, and the influence of other individuals is not taken into account; in the second, the genotype A is not affected until all individuals of the genotype B have died. However, the model of soft selection seems to be more realistic. One can hardly imagine the survival and reproduction of an individual within a group of plants or animals competing for limiting resources to be independent of the number and genotype of the other individuals within this group.

Note that the same two models can be applied in the case of different age classes instead of different genotypes. It is well known that the members of the weaker age classes (i.e., young and senile individuals) are the first to be affected by the shortage of resources, so that their mortality increases much more than that of the remaining members of the population. Field data concerning differences in mortality among genotypes are not as abundant as data on these differences among age groups, but there is no good reason why the same could not apply to different genotypes.

Soft selection seems to be a very important and straightforward concept, but it suffers from the same shortcoming as the models presented in sections 2.1 and 2.2: the process is not described by a single continuous equation, and therefore it is more difficult to handle mathematically. Nevertheless, some possibilities of a mathematical description of this process do exist. Following Wallace (1981), I shall present here a simple model of soft selection against recessive homozygotes, not only to make this concept more familiar, but also to show its consequences as compared to hard selection.

Soft selection can take place if the population size N is above the size, say k, at which it can be supported by the available resources. Let the coefficient of soft selection s be defined by the equation

$$s = (\mathcal{N} - k)/\mathcal{N}. \tag{4.7}$$

This coefficient does not describe the selection against a particular genotype, but it defines either the intensity of density-dependent reduction in survival, or reproduction, or both. When describing the selection against the recessive homozygotes aa occurring with frequency q^2, two separate cases should be considered. First, when the genotype aa is so frequent that $q^2 > s$, then the selection against this genotype is partially density-dependent, and the coefficient of hard selection s_{aa} to the model of soft selection is given by

$$s_{aa} = s/q^2. \tag{4.8}$$

Second, at low frequency of the genotype aa such that $q^2 < s$, there is a complete elimination of this genotype, which means that $s_{aa} = 1$. Note that the frequency dependence of soft selection means stronger selection against the genotype aa when this genotype is less frequent. This outcome is opposite to most cases of classic frequency-dependent selection, which acts against more frequent genotypes.

Applying the coefficient of hard selection s_{aa} to the model of soft selection makes it possible to calculate changes in the frequency q of the recessive allele a as a function of its frequency (Figure 4.3) for a given coefficient of soft selection s. These changes can be then compared with the classic model of hard selection, by assuming that $s_{aa} = s$. Although these two coefficients differ, since s refers to the entire population while s_{aa} refers to the genotype aa only, such a comparison allows us to show the effectiveness of soft selection in relation to hard selection. The mathematical description of the former seems to be less simple, but ecologically more realistic, and it uses a smaller number of parameters: what is really required is a value of s

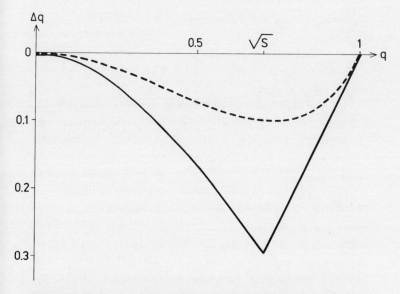

FIGURE 4.3. Changes in the frequency q of a recessive allele as the function of its frequency for the models of hard (broken line) and soft (solid line) selection. A selection coefficient $s = 0.5$ was assumed. In the hard selection model, s is the proportion removed from all the recessive homozygotes aa, whereas in the soft selection model, s is the proportion removed from the entire population.

that relates population size to the amount of available resources and to the abilities of genotypes to survive when resources are limited.

Look again at diagram (2) in Figure 4.2. Let the horizontal axis denote the ability of an individual to survive and reproduce, which in terms of the model from Chapter 2 can be measured in individual resource intake y. According to the concept of soft selection as presented here, the resource share y of each individual of genotype B is lower than the share y of any individual of genotype A. This is an unrealistic situation, in which the success of an individual is determined solely by the genotype in question. What has been said about the genotypes applies also to different age classes. It seems obvious that although the resource shortage is disproportionately more detrimental to the weaker age group, it also affects some individuals of the stronger age group. This is because neither age nor genotype is the only source of variation in individual ability to survive and reproduce. A more realistic model of soft selection should consider this variation.

A graphic model that explicitly includes this variation is given in Figure 4.2 (3). The quantitative character, plotted on the horizontal axis, is the individuals' ability to survive. It is assumed that the two genotypes differ in the mean values of their survival ability, and that their distributions overlap. Individuals that fall below a certain critical value die, and since the distribution of genotype B is closer to zero than that of genotype A, more individuals of genotype B die. Increased density, which causes additional mortality, moves the critical value upward, increasing mortality, but this additional mortality affects genotype B more than genotype A. This is yet another interpretation of the mortalities m_A and m_B mentioned above, one that does not require the assumption that m_A is an accidental mortality common to both genotypes and that gives yet another prediction concerning the distribution of additional mortality m_D among genotypes A and B. To make this prediction, a knowledge of the distribution of the ability to survive is required for both genotypes separately. Such knowledge can be obtained by studying survival within a group of individuals at different densities or at different concentrations of an adverse factor.

The concept of soft selection, as described above in both the second and the third model, shows that hereditary traits are the agents that place, or more precisely rank, individuals along the gradient of ecological success. But the concept of soft selection requires not only a ranking of individuals, but also a shortage of resources, which would cause either death, or reduced reproduction, or both, in those of lower ranking. This concept may appear mathematically messy, but in essence it is very close to the Darwinian idea of natural selection, and it is much less abstract and more realistic than the models developed by classical population genetics. The concept of soft selection corresponds very closely to the model outlined in Chapter 2, which requires the ranking of individuals according to their resource intake. For this model, the genetic variation among individuals is one of the factors determining individual rank, and, along with other sources of individual variability, it may influence the dynamics of populations.

4.3. GENETIC DETERMINATION OF INDIVIDUAL SUCCESS IN THE ECOLOGICAL WORLD

Placing hereditary variation along a single gradient of ecological success seems to run against established trends in population genetics and evolutionary theory. Contemporary population genetics views natural populations as genetically very variable (Lewontin 1974), this variation being maintained by different selection pressures in both space and time, as well as by all kinds of frequency-dependent selection against more frequent genotypes. The image of genetic variation is more like a mosaic in a multidimensional space than like a gradient along a single axis.

I do not think these two concepts exclude each other; they only represent different points of view on the same process. The main concern of population genetics is the study of hereditary variation and its maintenance against the forces of selection and genetic drift. For this reason, population geneticists have to look for spatial and temporal heterogeneity of habitats and for the reaction of different genotypes to this heterogeneity. Such heterogeneity may also be important in ecological processes, and it will be discussed later, in section 8.2 and in Chapter 10. However, for the basic model of population dynamics, in a given place and at a given time, some genotypes are adapted better than others to the acquisition of resources, survival, and reproduction, and therefore they can theoretically be placed along a single gradient of ecological success. Their position may change in time and space, not only due to changes in environmental conditions, but also because of their changing frequencies. I shall ignore these phenomena here. My main concern arising from the models presented in Chapter 2 is to answer the question of whether natural populations are variable enough to be stable.

As shown in Chapter 3, the variation of individual success, as determined by body size, can be a by-product of growth processes under limited amounts of resources. Is genetic variability also important for the existence of unequal resource partitioning and unequal individual success? There are two phenomena that leave room for doubt, and for this reason they deserve a short discussion. These are (1) the low heritability of important fitness characters, and (2) the higher morphological variability of homozygotes than of heterozygotes.

Data on heritability of different traits among domesticated animals show that traits that seem strongly correlated with fitness, such as natality, weight, and milk production, exhibit lower heritabilities than the traits that seem neutral, like coat color for example (Falconer 1981). This is explained by the strong selection pressure for the maximum values of the former traits. This pressure has exhausted the traits' genetic variability, so that it is mainly the environment, not heredity, that determines the fitness of individuals. If the same holds true for natural populations, then genetic variation is of minor importance in the determination of individual success, and consequently it has little effect on population stability.

Recent studies of heritability carried out within natural populations seem to confirm this view. Gustafsson (1985) has studied 12 different traits in a natural population of the collared flycatcher *Ficedula albicollis*. He has determined both the heritability of these 12 traits and their correlation with the lifetime reproductive success of the flycatchers under natural conditions. These data confirm the concept of the low heritability of important fitness traits. The variance of the life span and of the number of fledged young explains 42.1 percent of the variance in lifetime reproductive success among males and 26.3 percent among females, while the heritability of these traits is not significantly above zero. Other characters such as beak length or tail length, exhibit a heritability close to 50 percent and, at the same time, a very low correlation with lifetime reproductive success.

On the other hand, a weak correlation of a trait with lifetime reproductive success does not imply a weak correlation with survival and reproductive success during one season. If large beaks are advantageous in one year and small beaks in another, then beak size can strongly influence individual success in one year without influencing the lifetime reproductive success estimated for several years. Variability in space may exert the same effect that variability in time does: large beaks can be a great advantage in one place and a disadvantage in another, but on average they are of little influence.

The mechanisms of artificial selection are different from those of natural selection. A breeder can separate each character, as long as there are no pleiotropic effects, and he can apply strong and stable selection pressure for several generations. In nature we may expect selection pressure to change from season to season, and we may also expect strong correlations between different characters that have nothing to do with pleiotropy. Take for example the number of eggs laid, a character that can be selected for by a breeder, simply by choosing for further reproduction those individuals laying more eggs, irrespective of their other features, like plumage coloration. In the field a female with inappropriate plumage may have to spend more time avoiding predators and consequently may get less food and lay fewer eggs. Therefore, it is theoretically possible that a bird with the inappropriate plumage would have a low clutch size. This imaginary example is mentioned here not to question the value of estimates of clutch size heritability, but to indicate the difficulties inherent in estimating the importance of genetic variation. We simply do not control the everyday economy of plants and animals in nature, and therefore we cannot claim that some characters exhibiting high heritability are of minor importance for reproduction and survival in a given time and place. Thus, it can be concluded that genetic variation may be an important and sufficient source of individual variation in reproductive success, even if there is low heritability of the traits strongly correlated with fitness.

Another phenomenon that can be used to question the importance of genetic background in determining the phenotypic variability among individuals is the higher morphological variability among homozygotes. This phenomenon can be seen in inbred lines, which represent a high degree of homozygosity and in which the variation of many different traits is much higher than among hybrids. For example, the coefficient of variation of body weights of mice and rats, as well as that of wing lengths of *Drosophila melanogaster*, is about 50 percent higher for the inbred lines than for hybrids (Falconer 1981). The application of electrophoresis for the study of genetic variation shows that this is also a natural phenomenon occurring in the field among several species of plants, and among both invertebrate and vertebrate animals. For example, Fleischer et al. (1983) report higher morphological variance within those populations of the house sparrow *Passer domesticus* that exhibit lower allozyme homozygosity.

The correlation between heterozygosity and low morphological variability was explained by the stronger developmental homeostasis of heterozygotes (Lerner 1954). This concept is also used to explain the higher degree of morphological asymmetry among homozygotes. Whether developmental homeostasis is a sufficient explanation for this phenomenon is still an open question. Without attempting to answer this question definitely, I would like to point out that on the basis of the empirical data and the theoretical model presented in Chapter 3, an increased variation among individuals living under unfavorable conditions is to be expected. If we accept that homozygotes and inbred individuals usually exhibit lower adaptation to their environment, we also accept that they can exhibit an increased variation similar to that of individuals that live under unfavorable conditions. Although genetic variation may be one source of variation of individual success, the data on variability in inbred lines show that genetic variation is not a *necessary* condition for variation in individual success.

The importance of hereditary variation was shown during competition experiments among different varieties of the annual plant *Phlox drummondii* (Heywood and Levin 1984), in which the variance of individual plant biomass could be attributed both to the variety and to individual variation within varieties. These experiments have shown that although most of the variance, from 67 percent up to 100 percent, resulted from the differences among individuals within varieties, the remainder was due to differences among varieties. Therefore, although genetic variation may be one of the reasons for the existence of individual differences among competing individuals, we may still expect large phenotypic variance among genetically similar individuals, if for some reason the genetic variance decreases. Consequently, there is no reason to expect low population stability due to genetic uniformity.

Age and
Overlapping Generations

5.1. AGE-DEPENDENT INDIVIDUAL SUCCESS

If a population can be divided into groups that differ in their probabilities of survival and abilities to reproduce, then in order to predict the dynamics of the entire population, it is sensible to identify these groups. The profits gained by many insurance companies that base their predictions on age-dependent mortalities suggest that ecologists should follow them in identifying different age groups and their properties. On the other hand, one must realize that there are some features of natural populations that make it much more difficult to apply age-dependent models to them than it is to apply such models to human populations in developed countries.

First, identification of an individual's age is usually a problem. We must either mark the individual at birth or use various other methods of age determination that often can be applied only to dead individuals. Then we must determine the probability of survival and the fecundity of females within each age group, a task relatively easy to accomplish in the laboratory, but extremely difficult in the field. When estimating the survival within each age group, ecologists face the same problem that population geneticists do, namely, that of estimating a proportion from a sample. This problem has been nicely summarized by the following passage from Lewontin (1974, p. 268): "No technological breakthrough can alter the fact that under ideal conditions the standard error of a proportion is never smaller than $\sqrt{pq/N}$, so that if we want to be 95 percent sure that we know some proportion within 2 percent of its true value, we must count 10,000 cases." When facing such difficulties, it seems reasonable to weigh the cost of applying age-dependent estimates against the gains from more precise predictions.

Second, it is generally overlooked that age-dependent survival and reproduction is also density- and frequency-dependent, as described in section 4.1. Age-dependent mortality is obviously "soft" mortality; in other words, it is higher at higher population densities in relation to available resources, and an increase in mortality is disproportional among different age groups: the weaker age groups are the first to be affected in the same way that weaker genotypes are affected by soft selection. For this reason, life tables that list age-dependent survival for a given density and given environmental conditions do not enable us to predict age-dependent survivorship at higher or lower densities. There is a tendency among ecologists to think of a life table as a characteristic of a species, but although this may be true for a population kept under optimal conditions where density has no adverse effect, it is of little use for predicting age-dependent survival at high densities or under deteriorating conditions. The models that assume that an increase in density results in proportional increases of mortality in each age group (e.g., Poole 1974) are of theoretical interest, but they are unable to generate predictions concerning population dynamics. The reason that it is impossible to predict survival from constant life tables is simple: the proportions of surviving individuals do not indicate the mechanisms of survival; they describe only the net result of many different processes—not only aging, but also body growth, the quality of the local habitat, and the relation of the individuals of a given age group to other individuals of the same or of different age groups. For populations under optimal conditions, or close to optimal conditions, such as human populations in developed countries, age is a good predictor of the fate of an individual, but for others, especially plant populations, this may not be the case.

Recently, it has been recognized that for many plants and animals the body weight of an individual is a much better predictor of its fate than is its age (Werner 1975; Bacon 1982; Hughes 1984). To divide populations into size classes instead of age classes seems very sensible for plants and sedentary animals that differ considerably in size and whose age is difficult to determine. Size may also play an important role in other animal groups. For example, Bacon (1982) suggests that the individual size of the mute swan, *Cygnus olor*, is an important feature determining the dominance hierarchy, and he therefore, postulates the application of size distribution rather than age distribution for predicting the population dynamics of this species.

The importance of the size as compared to the age of an individual is hard to assess, because almost all population ecologists consider either size only or age only. The data collected by Werner (1975) on the survival and reproduction of the semelparous plant common teasel, *Dipsacus fullonum*, are a rare example of empirical results that have considered simultaneously both age and size of the members of a single population. Werner studied the proportions of plants of given age and size classes surviving to the next season and flowering in it (Figure 5.1). These proportions show that the size of an individual plant is much more important than its age for predicting the probability of its reproduction: except for a few three-year-old individuals that were able to reproduce despite their rather small size, the diameter of the plant rosette must be above 19 centimeters to make flowering possible in the next year. On the other hand, the claim that the age of an individual is of no importance for its fate is unjustified. Size is to a certain degree correlated with age, so that the largest size classes are not represented among the youngest individuals and the smallest sizes are not represented among the oldest ones. Figure 5.1 clearly shows that age is also an important determinant of individual fate: mortality within the class of largest individuals occurs only

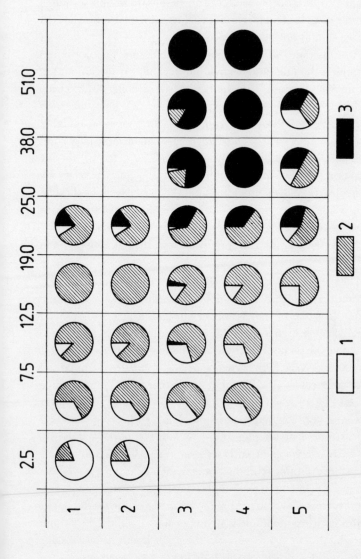

FIGURE 5.1. The empirically determined probabilities of (1) dying, (2) remaining vegetative, and (3) flowering in common teasel as determined by age (1 to 5 years, given in the first column) and the size of the rosette (in centimeters, given in the first line) in the preceding year. Data from Hubbell and Werner (1979)

among the oldest ones, and there are clear differences in flowering among individuals of the same size but of a different age. What can be said on the basis of these data is that although the size of an individual cannot be ignored in predicting its fate, its age should not be ignored either.

It seems that the relative importance of age and size in determining the fate of an individual depends on the amount of within age variation in the individual size. From what has been said in Chapter 3, we may expect higher individual variation in body size under worse environmental conditions and at higher densities. Under such circumstances, size can be more important than age, whereas under optimum conditions, the within age variation of body size can be quite low, allowing age to be the main determinant of individual fate. When studying the survival of a cohort under optimum conditions in a laboratory, with relatively low within age variance, one may get a biased picture, with age seeming much more important than it actually is in the fields.

There is another aspect of age that must be taken into account. Reproductive success is age-dependent; therefore, its variation within a group of individuals of all ages in a given season is usually much higher than the variation in individuals lifetime reproductive success. Nevertheless, lifetime reproductive success may also be very high. When studying a red deer population, Clutton-Brock et al. (1982) found that the lifetime reproductive success of red deer ranges from 0 to 13 offspring surviving at least one year for hinds, and from 0 to 24 offspring surviving at least one year for stags. This variation is partly caused by the early mortality of both sexes, and since some individuals die before attaining reproductive age, their lifetime reproductive success is equal to zero. For the determination of population stability from season to season, the survival and the reproduction of the entire population is of importance, while the lifetime reproductive success is not. On the other hand, if we consider an optimum strategy for an individual in a given season, for example whether to reproduce or to refrain from reproduction,

whether to migrate or to stay, it is the future lifetime reproductive success that matters. Up to now, I have considered an
individual organism as a machine built to convert resources into
progeny. When introducing age and overlapping generations
into a population of individuals with unequal resource partitioning, we must consider that an individual maximizes its lifetime reproductive success, not its seasonal reproductive success.
This fact may have enormous consequences for population dynamics, especially when the social position of high-ranking individuals is relatively stable from one season to the next, while
the position of young individuals is low. Under such circumstances, we may expect that at high population density in relation to available resources, not only juvenile but also adult
individuals may refrain from reproduction in order to reproduce
later when the density decreases. This may look like self-regulation of population density, evolved by group selection, but it
is only a by-product of unequal resource partitioning and overlapping generations. Since this is an important problem, it will
be discussed in more detail in section 7.3.

5.2. DISTINCT LIFE STAGES
WITHIN A POPULATION

Before discussing further the effect of age on resource partitioning and stability, a short account of a special kind of effect
of age should be given, namely that due to the existence of
different life stages or, according to Wilbur's (1980) terminol-
ogy, complex life cycles. A population is usually defined as a
group of all individuals of a given species in a given area, and
although the existence of some very peculiar life stages in many
species of animals is a well-recognized fact, there is a strong
tendency to think about a population in terms of single unit.
On the other hand, when competition is considered, the caterpillars of a butterfly species are one separate competing group,
while the adult butterflies form a quite distinct competing group,

with no competitive relation to caterpillars. In some species, this distinction between different age classes is not so deeply pronounced—for example, among young and adult individuals of land snails, which overlap in the kinds of resources used. Similar phenomena have been discussed by Werner and Gilliam (1984), who have recently given an account of the ecological consequences of ontogenetic growth and variation in body size. Other complications are also possible: for instance, the breeding pairs in a bird population form one competing group. The progeny of each pair within the nest form a separate competing group. The nestlings compete directly with their sibs in the nest and indirectly through their parents with nestlings in other nests.

There are complications in the definition of competing groups, but if such groups exist, they should not be ignored, and the process of competition within them ought to be studied. A hypothetical example of how population stability is determined by the stability of competing groups made up of different life stages is presented by the graphic model in Figure 5.2. This model is an extension of the model of population stability given in Figure 2.5. We may assume the first stage to be the larval one from hatching to pupation, the second to be from pupation to reproduction. Reproduction itself, after which the individuals of this generation die, is presented on the lowest diagram (R), which is simply a plot of the number of reproducing individuals multiplied by the average number of eggs produced per capita. In this particular example, the larvae are assumed to exhibit contest competition, the imagines scramble competition. The stability of the population as determined by different stages depends not only on the shape of the functions determining the density at the end of the stage $N(t, s + 1)$ in relation to the density at the beginning of the stage $N(t, s)$, but also on the position of these functions relative to each other. This particular example of the model produces small oscillations with a period of two generations, but we can well imagine an example of a stable population or of a population subject to frequent extinctions. Note that some stages can control the stability of the

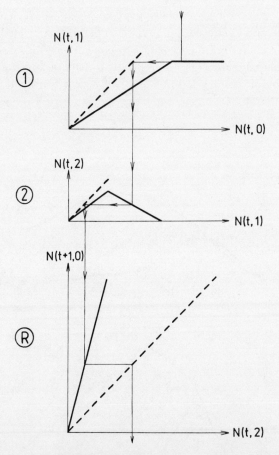

FIGURE 5.2. A theoretical example of the analysis of population stability during two life stages s within a single generation t, followed by reproduction (R). Diagrams represent population densities at the end of the stage $N(t, s+1)$ as a function of the densities at the beginning of the stage $N(t, s)$. The broken lines, which have a slope of 45 degrees, are used to transfer the actual value of $N(t, s+1)$ from horizontal to vertical axes, as shown in Figure 2.5. The lines connecting all three diagrams represent changes in population density.

entire population, especially those stages at which density is considerably reduced by contest competition. A related theoretical analysis of the dynamics of populations, with nonoverlapping generations encountering two different seasons and with the logistic assumption applied, had been given by Kot and Schaffer (1984).

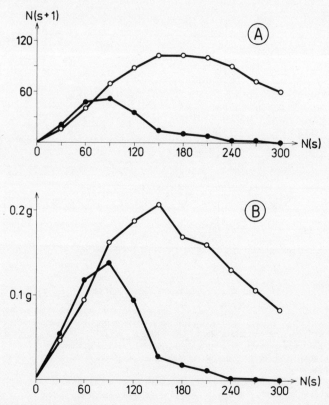

FIGURE 5.3. (A) Number $N(s+1)$ of individuals of *Tribolium confusum* beetles surviving to adulthood, as determined by the number $N(s)$ of eggs per gram of medium. Black circles refer to experiments in vials with no possibility of finding a safe pupation site; open circles refer to experiments in which larvae could migrate and pupate outside without being endangered by their less-developed companions. (B) The same data expressed as total biomass of pupae for each egg density. From Laskowski (1986).

When considering a single life stage as a competing group, we take into account the number of individuals at the beginning and at the end of the stage. These two numbers are related by survival only in a nonreproductive stage. In such a case, the ability of an individual to reproduce can be determined indirectly. For example, individuals that attained larger body size during the larval stage will be able to produce more offspring than those with a smaller body size. Even though it may be difficult to relate body size to future reproductive success, the differences among individuals at the end of a given stage should not be overlooked. For example, the outcome of competition among the larvae of *Tribolium confusum* can be presented as a function relating the number of pupae to the number of eggs, or as another function relating the biomass of pupae to the number of eggs. The example given in Figure 5.3, based on a study by Laskowski (1986), shows how the migration of the last-instar larvae out of the vials containing their less-developed companions affects the development and mortality of the entire group. The net result of such migration is the more efficient use of resources and the different shape of the function relating the densities of pupae to the densities of eggs. Experiments of this sort allow for a more precise description of the factors affecting population stability, as compared to the study of the entire population composed of individuals in all life stages. Even if such a study does not tell us whether the population is stable and persistent, it helps to identify the factors that affect population stability.

5.3. SIMPLE EXTENSION OF THE MODELS OF POPULATION DYNAMICS TO OVERLAPPING GENERATIONS

The model presented in Chapter 2 describes populations with nonoverlapping generations in discrete time units. Here I would like to introduce overlapping generations into this model. For the formal description of populations divided into age groups or size groups, one can apply the Leslie matrix model, which is commonly used for discrete age groups (Leslie 1945), and which was recently used for discrete size groups by Bacon (1982) and Hughes (1984). Considering both age and size simultaneously is a difficult task, especially if we allow for the complications arising from the existence of "soft" mortality as discussed in section 4.1. It seems that the simplest and most reasonable approach to the problem of age structure is to assume that age, like hereditary variation, is one of the agents responsible for unequal resource partitioning.

It is very difficult, however, to include age as a factor determining unequal resource partitioning. Older individuals are usually stronger, and get larger shares of resources than both the younger and the oldest ones. The presence of the oldest, senile individuals causes a real complication, because their fates cannot be predicted from their sizes alone. On the other hand, in the real world, where competition among individuals is strong and mortality is high, the group of senile individuals is small and its effect can be ignored.

An interesting review on the importance of age in natural populations of the bank vole *Clethrionomys glareolus* was given by Gliwicz (1979). She collected data on the sizes of home ranges of males as well as on the trapability and spatial distribution of both sexes for five different age groups, the oldest of which was made up of individuals that had survived the winter. These data clearly show the correlation between the age and the social status of an individual within the population. The older individuals occupy larger home ranges in better habitats, and they do not lose their high status during a given season.

In order to extend the models presented in section 2.1 to the case of overlapping generations, we can redefine the parameter m, which was defined earlier as the maintenance cost from birth to the time of reproduction. It m is redefined as the cost of maintenance during the period of one year, and y as individual resource intake during one year, then an individual that gets

$y < m$ resources dies before the next season, while an individual that gets $y > m$ resources produces $(y - m)h$ offspring and survives to the next season.

Version 1 of the model presented in section 2.2 by equations (2.3) and (2.5) changes into the following form: for $N(t) \leq V/a$, for which $y = a$,

$$N(t + 1) = N(t) + N(t)(a - m)h, \tag{5.1}$$

where a denotes the maximum amount of resources an individual can take during one season, and h is a conversion coefficient of resources into progeny. For $V/a < N(t) \leq V/m$,

$$N(t + 1) = N(t) + [V - mN(t)]h, \tag{5.2}$$

while for $N(t) \geq V/m$,

$$N(t + 1) = 0. \tag{5.3}$$

The last equation is the consequence of perfectly equal resource partitioning; therefore, if each individual gets $y = V/N < m$ units of resources, none can either reproduce or survive to the next season.

The relationship between $N(t + 1)$ and $N(t)$, as defined by equations (5.1) through (5.3), is represented in Figure 5.4 for three different cases: (A) $hm > 2$, (B) $1 < hm < 2$, and (C) $hm < 1$. The equilibrium point is at $N_e = V/m$. For case (C), in which $m < 1/h$, which means that the maintenance cost for a whole year is lower than the cost of producing one offspring, any population decrease will result in a slow return to this point, without oscillations. For cases (A) and (B) we may expect, respectively, divergent and convergent oscillations. The point of equilibrium N_e cannot be regarded as stable, because $N(t = 1)$ as the function of $N(t)$ is not continuous at this point. Formally we can imagine a population growing to the point at which each individual takes the same minimum share of resources $y = m$. This theoretically allows all individuals to survive to the next season or to produce enough progeny to replace dying

97

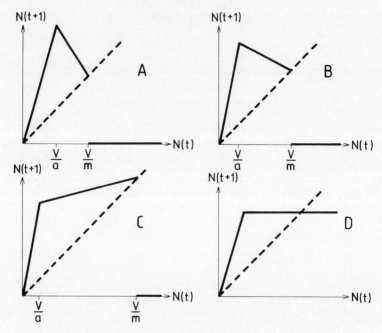

FIGURE 5.4. $\mathcal{N}(t+1)$ as a function of $\mathcal{N}(t)$ for populations with overlapping generations. Diagrams (A), (B), and (C) represent three cases (explained in the text) of version 1 with equal resource partitioning; diagram (D) represents version 4 with unequal resource partitioning.

individuals. On the other hand, if there is any temporal variation in the amount of resources, the equilibrium at point $\mathcal{N}(t) = V/m$ is impossible, because any random drop in the amount of resources results in extinction of the entire population. Such a strange theoretical prediction is the result of the very unrealistic assumptions that all members of the population take identical resource shares irrespective of their age and that they react identically to the shortage of resources.

With modifications introduced by overlapping generations, version 2 of the model from section 2.1 reduces to version 1, and version 3 to version 4. This last one, which assumes that resource intake y is either a maximum possible amount or zero, so that $y = a$ for V/a individuals and $y = 0$ for remaining ones, yields for $N(t) > V/a$

$$N(t+1) = (V/a)[1 + (a - m)h], \qquad (5.4)$$

while for $N(t) \leq V/a$, the equation is the same as (5.1). This implies stable and persistent systems, as shown in Figure 5.4(D).

These two models, although very simple, allow us to extend the concept of population stability in relation to unequal resource partitioning to overlapping generations. The equations describing population dynamics differ slightly from those for nonoverlapping generations; for version 1 the system is even less stable, but the relation between stability and resource partitioning holds for both overlapping and nonoverlapping generations.

Considering that in this case there are individuals of different age groups competing with one another in populations with overlapping generations, one may expect much larger differences among individuals than there are in groups composed of individuals of the same age. One may also expect older individuals to be able to secure part of the resources for themselves, and for this reason version 1 of equal resource partitioning is, for overlapping generations, mainly of theoretical interest. In the real world, extinction of the entire population, as predicted by version 1, is very unlikely, and one should instead expect the dynamics of populations with overlapping generations to be close to version 4.

Empirical data that can serve to illustrate population stability for overlapping generations have been reported, for example by Thomas et al. (1980) for 27 species of *Drosophila*, studied at two different temperatures, 19° and 25°C. The females of each species were allowed to oviposit for seven days at a wide range of densities $N(t)$, and then the number of females that survived these seven days, as well as the total number of their adult progeny, were summed up to represent population density $N(t+1)$ in the next generation. This is clearly an experimental counterpart of populations with overlapping generations, in which there are two generations exposed to different sources of mortality. The survival of larvae from eggs to adulthood has little relation to the survival of adult females during the seven days of egg-laying. The results of this experiment are given by 58 different sets of final densities $N(t+1)$, as determined by initial densities $N(t)$. In most cases, $N(t+1)$ as a function of $N(t)$ grows at a decreasing rate, and then either remains constant or reaches a maximum followed by a rather slow decrease. Only in 3 out of 58 sets does $N(t+1)$ eventually fall to zero, which implies the possibility of population extinction due to over-crowding. In the remaining sets no possibility of such an extinction exists. Since the absolute value of the derivative of the function $N(t+1) = f[N(t)]$ in its decreasing part is smaller than unity for all 58 sets of data, we can expect equilibrium points to be stable for all these sets. The authors suggest that the stability of these experimental populations of *Drosophila* can be attributed to the action of group selection, in which less stable groups are selected against due to violent oscillation resulting in group extinction. Putting aside the theoretical difficulties of accepting group selection, I see no need to invoke this kind of selection to explain population stability as expressed by these data. The existence of individual differences among *Drosophila* larvae, as well as the differences between two generations, is a sufficient explanation of the stability. The existence of these individual differences does not require an explanation by group selection either, since they can be regarded as a by-product of other biological processes, described in Chapter 3.

5.4. DISCRETE VERSUS CONTINUOUS MODELS OF POPULATION DYNAMICS

The fundamental assumption underlying all the models presented here is the discrete nature of the elements that constitute natural populations, so that important population properties determining stability cannot be understood if a single population is seen as an amorphous mass. There are obviously situations in which the behavior of such a mass very closely approximates the behavior of many small entities, as in the case of unlimited population growth; but when both reproduction and mortality occur, some individuals are more likely to die than others, some are more likely to reproduce than others, so that the population can hardly be treated as a shapeless mass.

The discrete nature of the units that constitute populations does not immediately imply the necessity of applying discrete time models of population dynamics; on the contrary, it is possible to apply continuous time models as well. I found discrete time models more appropriate for the description of the concept presented here for the following reasons.

First, there is a large class of habitats that are strictly seasonal; a given amount of resources is supplied to the population only once in the entire season, so that the discrete models are a good approximation of seasonality. These models are also a good approximation if there is a critical period of food availability for the population, so that the model of unequal resource partitioning and mortality as determined by this partitioning is the description of this particular period.

Second, when applying large time units (e.g., an entire season), we may well assume that between them all the individuals are able to exhaust all the available resources. For example, if there is a place for a plant individual to grow and to reproduce, within an area occupied by its population, there ought to be, during the entire season, at least one seed that is able to use this place to grow into a plant. Similarly, if there is an empty place for an animal, then one may expect that at least one individual will reach this place during the season. This may not be the case when we consider a short time interval. Unless the habitat is constantly homogenized, a change in one place, for example the death of an individual, does not imply that all other individuals are affected. Migratory and exploratory behavior of animals is usually limited in time, and therefore the reaction in one place to what has happened in another place cannot be immediate. I am afraid that the effect of such local changes cannot be described by introducing a simple time lag; what is really required is a model incorporating changes in both time and space. There are undoubtedly some populations of unicellular organisms that live on water-soluble resources. If the medium in which they live is constantly homogenized, as in some laboratory cultures, then any changes in the concentration of resources may immediately affect all the population members. Only under such circumstances can population models in continuous time be good approximations of real populations.

Third, during the time interval of an entire generation or an entire season, some well-defined large differences among individuals are established: some individuals survive and some do not; some reproduce and some do not. The application of shorter time units requires a more detailed description of the changes occurring among population members; we have to record, for example, that individuals grow or their body sizes diminish. More detailed models, based on much shorter time units, are of great importance for understanding what is really happening within the population, but they are immensely more complicated than the models presented here.

The arguments presented above in favor of the application of discrete time units in the model of population dynamics do not eliminate the important objection that the concept of stability as applied here is of importance for discrete models only. It is possible to develop continuous versions of the models presented here, and such an attempt has been made by Uchmanski (1983), but these versions differ considerably from the discrete ones presented here. To see clearly what the introduction of continuous time would mean, the models for overlapping generations would have to be studied, with time intervals decreased to infinity instead of being equal to one season.

With a decreasing time interval, the resource supply V for the entire population, individual resource share y and its maximum value a, as well as maintenance cost m per time interval, decrease proportionally. On the other hand, the cost of producing a single offspring $1/h$ remains the same, so that the average number of offspring produced per time interval decreases proportionally with a decrease of interval. Consequently, the relation between the maintenance cost m and the cost of producing a single offspring $1/h$ also changes. This implies that for very short time intervals, the important inequality (equation [2.20])

$$1/h > m,$$

which is the condition for the stability of version 1 of the model for nonoverlapping generations, is always fulfilled. As was pointed out in section 2.3, a low maintenance cost in relation to the cost of producing one offspring allows for stability even if resources are equally divided among individuals and even when generations do not overlap. A similar phenomenon occurs in the modified version for equal resource partitioning and overlapping generations (Figure 5.4[C]) on the left side of the point of equilibrium N_e. Therefore a decrease of time interval has a strong stabilizing effect, even if resources are equally divided among individuals. Nevertheless, for overlapping generations, the decrease of time interval does not theoretically allow for population stability, because even very short time intervals do not remove the discontinuity of $N(t + 1)$ as a function of $N(t)$. If for a given time interval the individual resource share $y = V/N$ is smaller than the maintenance cost m—in other words, if $N(t) > V/m$—the entire population becomes extinct.

But, it can be argued, the model presented above is far removed from reality, since we do not know of a single organism that might possibly behave in the way postulated by the model; usually, in a short time interval only a small fraction of population members dies of starvation, which results in an increase of resource intake of all remaining population members and allows the population to continue its existence. Consequently, it can be further argued, it is reasonable to assume that resource

requirements of the population and resource availability determine the fraction of individuals that die within a time interval.
Such an assumption can be applied, and under some circumstances it may give us a good prediction concerning population
dynamics. But on what premises do we base this approach? Is
it based on the behavior of population members and their relations to each other? The answer is definitely "no." We are
only exploiting an observation that more resources imply better
survival over the unit of time, and we base our model on a
strictly epiphenomenological description of the population's behavior: some individuals die, and more of them die if conditions
are worse, but we do not know which ones or for what reasons.

When one describes the dynamics of a population in discrete
time units by means of difference equations, one can obtain
either stable or unstable systems, depending on the assumption
concerning resource partitioning. But since the differential equations use infinitely short time units in which only a small fraction
of individuals can die or reproduce, the net result is similar to
that obtained in version 4 of the model of unequal resource
partitioning, i.e., some individuals die while other survive, and
we can get only stable systems. It can be said that differences
among population members are inherent in continuous time
population models. It accords with common knowledge that
within a short unit of time only a fraction of a population dies
or reproduces, not the entire population, and therefore we do
not require any special assumption about the mechanisms that
make this phenomenon possible.

On the other hand, if we really intend to understand the
stability of single-species populations and consequently the stability of larger systems composed of such populations, we must
avoid the implicit assumptions that result in stability. One such
implicit assumption is the existence of differences among population members that allow for only a fraction of them to die
or only a fraction of them to reproduce. Discrete time models
avoid this implicit assumption and allow for an explicit description of the mechanisms operating in populations subject to lim-

ited growth. Difference equations can be modified simply by shortening the time units, but differential equations cannot be modified into discrete equations by an equally simple method. This is the main reason why I have applied discrete time models here instead of continuous ones.

I will not insist that difference equations are the only possible tools for studying population dynamics. The models presented here are very simple; more precise ones would require describing a population of a single species by a system of growth equations, one for each individual. Such a system should also include a separate equation implicitly describing changes in the amount of available resources. The discrete models applied here are nothing more than a first approximation, allowing one to see the mechanisms underlying population stability. No such mechanisms are to be found in most of the continuous time models.

The Mechanism of
Contest Competition

As shown in Chapter 2, population stability results from two phenomena: unequal resource partitioning among individuals within a population and contest competition among them. Chapters 3 through 5 have discussed various mechanisms that lead to unequal resource partitioning, this chapter discusses the second phenomenon—contest competition.

Begon (1984), in an important paper on the consequences of individual differences within a single population, discussed the phenomenon of asymmetric competition, which according to his definition can be equivalent to contest competition as defined here. He has attributed the existence of asymmetric competition to various differences among individuals, but without clearly distinguishing between two phenomena: differences per se between individuals and the asymmetric relations due to these differences. I think, on the other hand, that contest or asymmetric competition is not a simple consequence of unequal resource partitioning, or of any other quantitative differences among individuals, and that it requires more detailed examination.

In section 6.1, contest competition is defined more closely as a particular type of interaction between members of a population; in section 6.2 it is described as a characteristic of a population, and an example of such competition acting in the field is given in section 6.3. Difficulties with the explicit formulation of the relation between scramble competition on the one hand and dominance hierarchies and territoriality on the other are discussed briefly in section 6.4, and section 6.5 presents the theoretical model of an arms-race approach as applied to contest competition. Further discussion concerning the evolution of behavior that leads to contest competition is presented in sections 7.1 and 8.1.

6.1. DEFINITION OF CONTEST COMPETITION

In section 2.5, contest competition was defined as a relationship between an individual and other members of the population, such that the resource intake of the individual in question is independent of the number of other weaker individuals. Using the symbols introduced in Chapter 2, contest competition means that an individual's resource intake y is a function of its rank x and the total amount V of resources, but is independent of population size N. This definition is derived from the model presented in Chapter 2, but, as shown below, it can also be applied to models other than this particular one.

The best illustration of the independence of individual resource intake y from the number of individuals of lower ranks is the competition for light among plants, and especially among forest trees. A large tree determines how much light penetrates through its canopy and reaches small trees or seedlings, but these small trees and seedlings, however numerous they might be, are unable to limit the amount of light reaching the canopy of the large tree. This simple asymmetric relation holds true for light only, since the competition among trees for nutrients and water may allow smaller individuals to influence the resource intake of the larger ones.

There are many examples of contest competition, as defined above, among animals. For example, the first solitary parasitoid larva to infect a host usually defeats any subsequent larvae that happen to find themselves in the same host (Hassell 1978). On the other hand, the mechanism of this defeat, like other mechanisms of competition among animals, is not as simple as in the case of one tree being shaded by another. For trees, the mechanism of competition can be clearly seeen and easily altered experimentally. For parasitoid larvae, the process of competition occurs within a host and is much more difficult to examine closely.

The system of contest competition for light does not require any special adaptation by either the large tress or the small ones. The shading of the soil surface by a large tree can be explained as a by-product of adaptation to the efficient use of light, while the limited growth and mortality of small trees and seedlings under the canopies of large trees can be regarded as a physical necessity. Since competition for light is a common phenomenon in plant communities, one can also expect contest competition and consequently the stability of plant populations to be a common phenomenon, or at least a more common one than among animals. Plant communities made of stable populations may exhibit higher stability than animal communities.

Contest competition is therefore based on an asymmetric relation: larger individuals can limit the resource intake of the smaller members of the population, but the latter are unable to limit the resource intake of the former. Consequently, scramble competition should be defined as a relation among population members, such that even a very small individual may alter the resource intake of all the remaining members of the population. Contest versus scramble competition can be illustrated by a diagram (Figure 6.1) in which the resource intakes $y(x)$ of two individuals are related to their body sizes $W(x)$. When describing scramble (S) competition, it is simplest to assume that there is a linear relationship between the resource share $y(x)$ of an individual and its total size, root surface, or some other measurement important for resource intake. If the larger individual takes a disproportionally larger share of resources (S/C)—i.e., if the relationship is not linear—this implies the action of an additional mechanism leading toward contest competition. Nevertheless, the competition should still be classified as scramble. This is because only full monopolization of resources, such that the larger individual takes as much as it

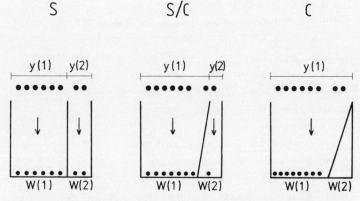

FIGURE 6.1. A simple representation of the mechanism of scramble (S) versus contest (C) competition. Sections $W(1)$ and $W(2)$ denote the sizes of two individuals. For scramble competition (S), the line separating these two individuals is exactly vertical, so that the food available (represented by the upper dots) is divided into two parts, $y(1)$ and $y(2)$ (shown by lower dots), linearly proportional to their sizes $W(1)$ and $W(2)$, respectively. Contest competition introduces a new element into the system, symbolized by the inclination of the vertical line. Available food is no longer divided proportionally, because the inclination of the vertical line gives the larger individual a disproportionally larger share (S/C) or deprives the smaller individual of its share altogether (C). According to the definitions of scramble and contest competition, the disproportional intake of resources without full monopolizatiion represents scramble competition, not contest competition.

requires, represents pure contest competition (C), according to the definition given above. Figure 6.1 depicts the special case in which the total available amount of resources is lower than the amount required by the larger individual. Contest competition does not necessarily require that all the resources should be taken by a larger individual; it only demands that the smaller individual should not be able to alter the resource intake of the larger one. Note that contest competition can be related to soft mortality (section 4.2): a stronger genotype is not affected by deteriorating conditions unless all individuals of a weaker genotype are eliminated.

Figure 6.1 helps to pose a question: What is the mechanism bringing about disproportional resource partitioning, up to its full monopolization by larger individuals? In the case of competition for light among forest trees, the answer to this question is straightforward, but for other limiting resources there is no simple answer. Figure 6.1 allows us to see that although asymmetry between individuals is a necessary condition for scramble competiton to occur, it is not a sufficient condition. Therefore, an asymmetric competition cannot be equated with contest competition.

It seems that the most important mechanisms of contest competition are social hierarchy and territorial behavior, disscussed in section 6.4. These mechanisms prevent the exploitation of resources by individuals on equal terms, because individuals fight for some prize other than resources, such as territory or social rank, and only after gaining this prizes are they able to secure resources for themselves. Can contest competition therefore be related to interference competition, and scramble to exploitative competition? Such relations were suggested by Nicholson (1954) and are worthy of closer examination.

Does interference among individuals always lead to contest competition? Larvae of *Tribolium confusum* kept in confined vials on a limited amount of food delay their pupation, which prevents them from being eaten by their less-developed companions (Dawson 1975). Cannibalism is an example of interference competition, but delay of pupation does not bring about a monopolization of resources by stronger individuals. As shown by data presented in section 5.2. (Figure 5.3), the opposite is true, so that the competition is closer to being of the scramble type in confined vials, where there is a threat of cannibalism, than in open vials. Furthermore, in the confined vials hardly any individuals survive at high density, which is a predicted result of scramble competition. The territorial behavior of Everglades pigmy sunfish (Rubenstein 1981b) described in section 6.4 is another example of competition with strong interference among individuals that cannot be classified as contest competition.

Interference is therefore not a sufficient condition for contest competition. Is it also not a necessary one? Among animals no example of contest competition without behavioral interference is known. A sort of social hierarchy occurs even among such primitive organisms as sedentary anthozoan coelenterates (Brace and Pavey 1978). It is an open question whether the competition for light among plants, which, as mentioned above, is the best example of contest competition, ought to be classified as exploitative or interference competition. Since taller plants interfere with the resource intake of the shorter ones, it may be concluded that interference is a necessary though insufficient conditions for contest competition to occur.

Competition for light among plants is a special case in which the competitors (trees) are unable to move, and the resource (light) flows to them from one direction, so that individuals in better positions (tall with large canopies) are able to get as many resources as they require, leaving the remaining resources for the individuals that are in worse positions. A similar system is theoretically possible in a population of predators that ambush along a path on which prey moves in one direction; this was mentioned in section 3.6, when the version of resource partitioning given by equation (3.11) was presented. This system raises questions: What prevents a predator from moving forward along the path, in order to take a better place supplied with more dense prey? To what extent is the mechanism of contest competition due to the action of individuals that secure a better position, and to what extent is it due to the action of those which seem to accept a worse position? Attempts to answer these questions are given in sections 6.3, 7.1, and 8.1.

6.2. POPULATION EFFECTS OF CONTEST COMPETITION

As shown in section 2.5, the most important effect of contest competition is the independence of population density $N(t+1)$ from the population density a generation ago $N(t)$, provided that population requirements are higher than the amount of available resources. The effect of scramble competition, on the other hand, is expressed by $N(t+1)$, which is a decreasing function of $N(t)$. This result is valid not only for the particular versions of the model presented in Chapter 2, but for any versions of this model. If both individual resource intake $y(x, V)$ and the minimum amount of resources necessary for survival m are independent of N, then the number of individuals k surviving to the time of reproduction, those for which $y > m$, and the total amount of resources $\sum(y-m)$ used for reproduction by the entire population are also independent of N.

The versions of the model presented in Chapter 2 include either dependence of y on N or independence of y from N, but they do not allow for a dependence of m on N. Such dependence can occur when the presence of other individuals causes the expenditure of additional energy. The larvae of *Tribolium* mentioned in section 6.1 are a good example; they prolong the duration of their larval stage in the presence of younger larvae, thus avoiding being cannibalized as pupae. It is rather difficult to find a population in which m is dependent on N while y is independent of it; therefore, the decision as to whether a situation in which m is dependent on N ought to be classified as scramble competition is mainly of semantic interest. However, since the result of this dependence makes $N(t+1)$ dependent on $N(t)$, the definition of scramble competition has to be extended to include cases in which the minimum maintenance cost m before reproduction depends on the population density N.

The effects of contest and scramble competition can be seen not only in two consecutive generations, but also within one generation, as shown by the number of survivors, k, as a function of the initial number of individuals, N (Figure 2.3). These effects may also be seen in two life stages within the same generation. Applying the graphic model of different life stages presented in section 5.2 (Figure 5.2), one can compare the number $N(s+1)$ of pupae as a function of the number $N(s)$ of larvae. Similarly, one can compare the number $N(s+1)$ of birds occupying territories as a function of the number $N(s)$ of contestants for these territories. The problem arises here how to interpret $N(s+1)$ when it happens to be an increasing function of $N(s)$. According to the model from Chapter 2, $N(t+1)$ is an increasing function of $N(t)$ for low values of $N(t)$ only, when neither kind of competition occurs. When competition does occur, then $N(t+1)$ is a decreasing function of $N(t)$ or independent of $N(t)$.

Consider the number $N(s+1)$ of birds occupying territories as a function of the number of contestants $N(s)$ for these territories. The fact that $N(s+1)$ is a linearly increasing function of $N(s)$ can be interpreted as evidence that no competition for territories is taking place. On the other hand, if we have the additional information that competition is taking place, then this can be evidence of scramble competition: if the number of territories does not increase with the number of birds trying to obtain them, then obviously some of them are able to monopolize the area, and their resource intake y is independent of the number N of birds in the entire population. An increase in the number of territories with the number of birds can be due to the lack of competition, but if we know from another source that there are more contestants for territories than places to establish territories, this shows that monopolization is impossible, that mean territory size decreases, and that at the stage of acquiring territories scramble competition occurs.

The phenomenon mentioned above can be exemplified by the egg-laying of flour beetles *Tribolium castaneum* (Łomnicki and Krawczyk 1980). Beetles of both sexes were placed in three different quantities in vials with four different amounts of medium (Figure 6.2). In the experimental vials the beetles were able to emigrate and had no possibility of subsequently returning to the vial; in the control vials, no emigration was possible. The number of eggs laid in the control vials over three days depended on both beetle density and the amount of medium, but it was independent of the number of beetles in the experimental vials. The number of eggs in the experimental vials, from which beetles were able to emigrate, was linearly related to the amount of

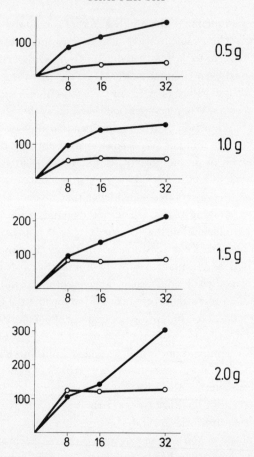

FIGURE 6.2. Number of eggs laid over three days, as a function of the number of beetles (8, 16, 32) placed in vials with four different amounts (0.5, 1.0, 1.5, 2.0 grams) of medium. In open vials from which beetles could emigrate (open circles), the number of eggs was independent of the number of beetles, whereas in the closed vials (black circles), the number of eggs was an increasing function of the number of beetles. Data from Łomnicki and Krawczyk (1980).

medium only, being very close to 59 eggs per gram of medium. One can expect that for vials with more medium and a smaller number of beetles, the number of eggs will be an increasing function of the number of beetles laying these eggs, even if emigration were allowed; for this particular experimental design, however, the desity at one stage (eggs) is independent of the density of the earlier stage (imago beetles). Therefore, if emigration is allowed, contest competition occurs, whereas without emigration the competition is of the scramble type.

The above example is far removed from commonly held views about contest competition. As mentioned earlier, contest competition among animals usually requires an interference among competing individuals. Very little is known about behavior and interference among flour beetles in the vials, but one can safely claim that there is much less interference when there is a possibility of emigration outside the vials than there is in closed vials where no such possibility exists. *Tribolium castaneum* is much more migratory than *Tribolium confusum* (Sokoloff 1977), and after laying eggs all the beetles leave the medium if they have such an opportunity. The experiments reported here show that in this particular design the beetles emigrate after laying about 59 eggs per gram of medium. It is still unclear why some females withdraw from the vials first, allowing others lay eggs, and therefore why, above the density of 59 eggs, further egg-laying is of lower advantage than the risk of emigration in search of a new place. One can imagine that such emigration would be advantageous if there was only one female in a vial, but for several unrelated females, staying in the same vial and cannibalizing eggs laid by others seems to be the better way of protecting their own offspring.

Whatever the advantage of emigration above a certain egg density may be, the experiments described here raise the question, mentioned already in section 6.1, of how important for the existence of contest competition is the behavior of those individuals which withdraw, and whether this behavior is adaptive. The control experiments described above show that the beetles that are forced to stay in the vials are physiologically able to lay more eggs and that they are not killed by their stronger companions, nor do they die of hunger. In spite of this, emigration outside the vial takes place, and this emigration makes contest competition possible.

115

6.3. THE CASE STUDY: COMPETITION AMONG GALL APHIDS *PEMPHIGUS BETAE*

There is a shortage of detailed and precise empirical data on the mechanism of contest competition among animals. Some recent studies in which detailed data can be found are those by Whitham (1978, 1980) on a gall-forming aphid, *Pemphigus betae*. Stem mothers are hatched from eggs that overwinter in deeply fissured bark; they then colonize young leaves in which galls are formed. The stem mother produces offspring viviparously and parthenogenetically; these offspring later colonize the secondary host. The stem mother's weight and her reproductive success depend on leaf size and the place on the leaf in which her gall is formed. A gall formed at the base of the leaf ensures the highest probability of survival of the stem mother, her highest weight, and her highest reproductive success. The females placed in more distal parts of the leaf exhibit higher mortality, lower weight, and lower reproductive success. Before the galls are formed, the females compete with each other for a better place on a leaf by kicking and shoving contests in which a larger female usually wins the superior basal position. The loser gets a worse, more distal position.

Whitham has analyzed the mean reproductive success of females in relation to the leaf size and number of aphids on that leaf. He was not concerned whether the competition among aphids was contest or scramble, but in his data (Whitham 1980, p. 460, fig. 6) one can find the average body weight and the average reproductive success of females placed at two different positions on 66 leaves of the size 12.3 square centimeters each, with two galls on each, and of females placed at three different positions on 54 leaves of the size 14.6 square centimeters, with three galls on each. Comparing these data with the data on mean body weights (Whitham 1980, p. 456, fig. 4) and their mean reproductive success (p. 454, fig. 3) on leaves of the same sizes with one gall only, one can find whether the presence of more distally placed females reduces the reproductive success of the female at the base of the leaf. These comparisons, which have been made for each leaf size separately, show that the presence of a stem mother at more distal places on the leaf does not decrease the weight and reproductive success of the stem mother placed closer to the leaf base. As shown by Whitham, the presence of more females diminishes the mean reproductive success, but this is due to the poor performance of more distally placed females, not due to a decrease in the performance of those at the base.

If the comparisons described above have been carried out correctly, then the rivalry among stem mothers of *Pemphigus betae* on a single leaf would be an example of ideal contest competition. The resource intake $y(x, V)$ of an individual is determined by its place x on a leaf and by the leaf size V, but is independent of the number of females of lower rank (i.e., at more distal positions on that leaf). When there are more females on a leaf, it is more difficult to find a place at the base of the leaf; therefore one can say that in contest competition, some dependence of resource intake on the number of other individuals is due to the shortage of places that makes an individual a high-ranking one. If an individual is of high rank, its resource intake does not depend on the number of individuals of lower rank, and therefore one can say that y is independent of population size N.

Contest competition among stem mothers as described here is not caused by any particular structure or physiology of the leaves. Since the products of photosynthesis flow from distal to basal parts of leaves, a decrease of resource intake y of an aphid àt the base due to the action of those more distally placed is theoretically possible. Aphids placed more distally can also injure the leaf, depleting in this way resources available at the base. It would be interesting to find out whether contest competition could persist at much higher densities of stem mothers on a leaf, provided that they would not have the possibility of emigrating outside this leaf.

This raises the question of whether the withdrawal of a smaller female to a more distal place on the leaf is due solely to the action of a larger one. One can imagine that the kicking and shoving between two females forces the weaker competitor to give way, but it is also possible that this kicking and shoving is a means of exchanging information about the competitors' relative strength. In the first case, the withdrawal would be a physical necessity for the weaker individual. In the second, the withdrawal of the weaker female would be an adaptation: she could stay as close to leaf base as possible despite being kicked and shoved, but for some reason she leaves the place already occupied by a stronger female. We can imagine the reason: an attempt to form its gall in the vicinity of a stronger female may end with the death of the weaker one. If so, then the withdrawal of a weaker individual to a worse place or even to another leaf is an adaptation that gives it better prospects.

If the withdrawal of weaker females is a behavioral adaptation—a choice made by these females—then the situation is similar to the case of egg-laying by *Tribolium*, described in section 6.2. In the case of *Tribolium*, we do not know the selective advantage of leaving the medium when egg density reaches a certain threshold. Here one can imagine the selective advantage of withdrawal in the presence of a stronger individual. The net result of these withdrawals is contest competition, as defined above.

6.4. SOCIAL HIERARCHY, TERRITORIALITY, AND CONTEST COMPETITION

Is either social hierarchy or territoriality a necessary and sufficient cause of contest competition among animals? In spite of a large amount of data about these two phenomena, this question is still open. It is very difficult to imagine strictly exploitative competition among organisms, which would lead to full monopolization of the resources by some of them, if others are able to move actively and choose a better place. Such monopolization requires some kind of behavioral interaction, as in the case of the aphids discussed above. One may suppose that some behavioral interactions among *Tribolium castaneum* beetles lead to contest competition in egg-laying when emigration is allowed, although neither territoriality nor social hierarchy is known in this species.

Theoretically, both social hierarchies and territoriality are means by which an individual secures some resources against the encroachment of others. Therefore either of these phenomena should result in contest competition. This is generally true, but some qualifications must be made.

To establish social hierarchies requires time, and at the beginning of these process, the monopolization of resources by dominant individuals can be very poor. Social hierarchies are well defined if there is a shortage of resources, in the same way that the variation in individual weights is much higher at low resource density (section 3.2). Inequality, which increases with the shortage of resources, can lead to the monopolization of resources and to contest competition, but it does not immediately imply contest competition. Dewsbury (1982) reviewed data on dominance hierarchies, mainly among mammals, and he concluded that high dominance rank is not always correlated with the frequency of copulation and the reproductive success of an individual.

A similar qualification should be made about territoriality. It is well known that territorial behavior sets an upper limit for the density of breeding individuals, a phenomenon that has been confirmed by immediate occupation of empty territories from which territory holders have been removed (Morse 1980). It does not mean that the density of territories is strictly determined by the resource density. Knapton and Krebs (1974) have shown that the colonization of empty territories by song sparrows *Melospiza melodia* gives different results, depending on the way the former territory owners are removed. The simultaneous removal of many birds leads to the simultaneous fighting of many newcomers for the space and results in the new division of this space and a higher density of newly established territories.

Individual resource intake y depends, in this case, not only on the amount of resources V, but also on the number N of individuals competing for these resources. If the removal of former territory holders is consecutive, it does not increase the number of territories per unit area, and sometimes it can even decrease it.

Most of the data on territorial behavior refer to birds studied in the field, where immigration and emigration was possible. Rubenstein (1981b) studied the territorial behavior of Everglades pigmy sunfish under laboratory conditions in closed tanks. He found that if the food for a group of fish is regularly supplied at the center of the tank, then some individuals defend territories at this center. If food is randomly distributed around the tank, territorial behavior disappears. If food is supplied at the center and, therefore, well-defined territorial behavior among the fish does occur, the net result of this behavior is not contest competition as defined here. Food intake and increase in body weight of the strongest individuals, which hold territories at the center, depend on the number of competitors. Furthermore, those fish which hold the best territories do not always take the highest amount of resources, and their gains in body weight are sometimes smaller than those of other individuals. There is obviously no monopolization of resources by territory holders. Why does territorial behavior in confined laboratory habitats not bring about contest competition? This is an open question, discussed in section 8.5.

6.5. EVOLUTIONARILY STABLE ARMS INVESTMENTS, OR HOW CONTEST COMPETITION CAN BE REGARDED AS A RESULT OF ARMS RACES

Up to now, those features of individuals which allow them to attain a higher rank and to win in competition—body weight for example—have been regarded as a by-product of body growth (Chapter 3) or as a by-product of the adoption of mixed strategies in an unpredictable environment (section 4.1). In this section, a theoretical model proposed by Parker (1983) is discussed, a model concerned with the evolution of the features designed to win in competition. This is the model of an evolutionarily stable strategy for arms races.

An individual that makes an investment that enables it to win a conflict bears a cost c of this investment, which results in lower survival, or lower reproduction, or both. This kind of game between individuals differs from the case of the war of attrition (Maynard Smith 1982) because it is an opponent-independent cost game; in other words, the cost of the conflict is determined by the arms level, which has to be prepared before the conflict takes place. Consequently, the cost does not depend on how well the opponent is armed or how long he is prepared to fight.

Assuming an investment level x, which involves a certain cost $c(x)$ and results in an arms level $R(x)$, no evolutionarily stable investment level x^* exists. This is because each individual that makes a higher investment will win the conflict. Parker (1983) has defined such a situation as a perfectly heritable arms level, because the arms level $R(x)$ is perfectly determined by the investment level x.

If one assumes that an investment level x does not determine the arms level $R(x)$, but rather the mean of the distribution of arms level $p(R)$, then an evolutionarily stable investment level is possible. Imagine an arms level determined by body size, which depends not only on the investment level x but also on many different environmental factors, so that an arms level $R(x)$ is not perfectly heritable. A higher investment level x moves the mean of the distribution $p(R)$ of the arms level toward higher values of R (Figure 6.3), but there are still some individuals that exhibit a very low arms level for other reasons. Individuals with an arms level above a threshold value T are able to survive and to leave progeny, and there are more such individuals if the investment level x is higher. On the other hand, a higher

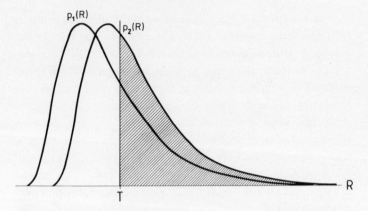

FIGURE 6.3. Two different distributions of arms levels $p_1(R)$ and $p_2(R)$, as determined by two different investment levels x_1 and x_2, respectively. Only those individuals whose arms level R is higher than a threshold value T are able to survive and to leave progeny.

investment x carries a higher cost $c(x)$, which diminishes the reproduction and survival of individuals in some other way. Since the higher investment level x does not make all individuals bearing the cost of these investments better than those adopting a lower investment level, an evolutionarily stable investment level is possible. Parker (1983) has presented several versions of this model, all of them allowing for an evolutionarily stable investment level x determining an evolutionarily stable distribution $p(R)$ of arms level.

This model does not require the assumption that an individual that loses the game acquires some other advantages at other times or in other places. Therefore, this can be a model of the game for survival, in which individuals with lower arms levels die without leaving progeny. In such a game for survival, an individual has no choice, since the choice was made earlier when a given investment level was chosen. On the other hand, this model does not need to be a game for survival, since the defeat of the loser may be a relative one, giving him other possibilities of surviving and leaving progeny. In such a case, it is advantageous for a weaker individual to give way in order to avoid an escalated fight with a stronger competitor.

Parker's (1983) model of an opponent-independent cost game explains how the relations between individuals that lead to contest competition can arise, but it does not solve the question of how the result of competition depends on the behavior of those individuals which are the losers in conflicts. This question will be further dealt with, and some appropriate evolutionarily stable strategy models presented, in sections 7.1 and 8.1.

Self-regulation of
Population Size

The models presented in the earlier chapters assume that population size at equilibrium is determined by the amount of available resources and how these resources are divided among individuals. It was explicitly assumed in the basic model presented in section 2.1 that the number of offspring produced by an individual is linearly related to its resource intake, and that an individual will die if it does not have sufficient resources.

In this chapter, I will discuss the possibility of the existence of populations in which available resources do not directly determine mortality and natality, and in which population density in relation to the amount of resources is lower than the maximum possible. While discussing this, I will ignore a possible reduction of population size by adverse physical factors and the action of other species: competitors, predators, parasites, or diseases. Therefore, population self-regulation is defined here as a phenomenon that maintains population density at a lower equilibrium level than the maximum possible for the given amount of resources, assuming that this level is not kept down by adverse external factors.

In this chapter, I will also refrain from considering more complicated mechanisms of the determination of population density, such as patchy resource distribution within a space or the differences in use and suitability of resources. Spatial heterogeneity and its consequences for population dynamics are discussed in Chapters 8 and 10. The relationship between the pattern of use and the suitability of different resources, and the importance of this relationship for maintaining population density below carrying capacity was recently discussed by Soberon (1986). The phenomena on which Soberon's concept is based—for example, the inability of some individuals to reach the most suitable resources—are important for population dynamics, but his approach is different from that advocated here. He considers the distribution of individuals among resources of different suitabilities as a population phenomenon, subject to statistical description, without a detailed analysis of what makes one individual choose a given kind of resource and how this relates to other differences among individuals.

The problem of self-regulation was widely debated by ecologists more than twenty years ago, especially after the publication of Wynne-Edwards's (1962) book, which postulated population self-regulation by means of various behavioral mechanisms. Wynne-Edwards proposed group selection as the evolutionary explanation of such self-regulation. Since then we have learned much more about group selection, which is now considered to be a rather weak evolutionary force (Maynard Smith 1964, 1976). Although Gilpin (1975) and Wilson (1980) have tried to defend the concept of group selection, it seems to be extremely difficult to show analytically how group selection might bring about population self-regulation (Łomnicki 1980a). Note that self-regulation implies soft selection against self-regulating genotypes, such that all self-regulating genotypes are removed before others are affected.

One problem related to that of self-regulation is a phenomenon that is very apparent in natural communities: a low density of herbivores in relation to their plant resources. Hairston et al. (1960) explained this by the action of predators that limit herbivore density. Van Valen (1973) called this problem "the Enigma of Balance" and presented it in the following passage: "How can it be that some species regulated even ultimately by food do not periodically greatly reduce their food supply by overeating?" (p. 31). In the same paper he wrote that "it is one of the most important well-defined problems of ecology" (p. 31).

Before "the Enigma of Balance" can be regarded as an important ecological problem that may have to be explained by population self-regulation, some simpler explanations must be considered. At least three of these are often overlooked by ecologists. First, to determine whether herbivores are food-limited or time-limited, and whether the presence of predators or of some adverse physical factors prevents herbivores from grazing in a particular place at a particular time, often requires separate studies. Second, many plants either have evolved resistance to herbivores or are useless as food for other reasons. Third, when determining the food requirements of herbivores, we either carry out laboratory experiments or infer these requirements from data on domesticated animals. It is quite likely that wild animals, which face a much harsher environment, require food of much higher quality, and therefore our assessments may overestimate the food supply available to herbivores.

In this chapter, the problem of population self-regulation is presented in a simple model as a conflict between two individuals (section 7.1). Experimental data on self-regulation are reviewed in section 7.2, and section 7.3 is an attempt to interpret these data.

7.1. SELF-REGULATION IN TERMS OF GAME THEORY

Is an adaptation possible that would make an individual give way to another, gain nothing, and lose all possibility of surviving and leaving progeny? As shown below, the answer depends on whether these two individuals are relatives. I will first attempt to answer this question for two unrelated individuals.

This is not the same problem as that in the well-known Hawk-Dove game, in which Dove can gain something by giving way. Here I will discuss an existential game, in which a defeat is absolute, with no possibility of survival or reproduction at another time or another place. The problem considered here is not solved by Parker's (1983) model of an opponent-independent cost game (section 6.5). In Parker's model, the withdrawal of the weaker opponent is assumed and is not regarded as an adaptive trait, the evolution of which should be explained by the model.

I will assume here an extreme version of the model of self-regulation, in which withdrawal seems to be most plausible. Imagine an organism that requires one undivided piece of a resource for survival and reproduction. The organism could be, for example, a larva of the grain weevil *Calandra granaria*, which requires one grain for its development and which dies when outside of this grain. The following assumptions are made here: (1) If there is one individual in a grain that individual will survive and reproduce. (2) If there are two individuals, neither of them will survive. (3) An individual that goes outside the grain will not survive. I will introduce two strategies that apply these assumptions: Hawk *H* always stays in the grain, and Dove *D* leaves the grain if another individual is present there. Which of these two strategies is an evolutionarily stable strategy (ESS)?

This model is very simple, but its conclusions are not always intuitively obvious. Two individuals in one grain can be considered as a randomly formed trait-group, as defined by Wilson (1980). Wilson postulated that altruistic behavior is more likely to occur among members of a smaller trait-group. The group of two is the smallest possible; therefore, if the theory of trait-group selection allows for self-regulation, it should allow for self-regulation in the model considered here. Imagine that Dove's strategy is an ESS. Each time the shortage of resources occurs, some individuals will give way to their neighbors, moving away or taking a much worse position. Such behavior would explain perfectly the evolution of territorial behavior and dominance hierarchy. This would also lead to contest competition, because of the monopolization of resources by one individual and withdrawal of another.

As the pay-off matrix for this game shows (Table 7.1), Dove strategy is not an ESS, but Hawk strategy is. It does not pay

TABLE 7.1. Pay-off matrix for the existential game between two individuals in a place that is able to support one individual only. Two strategies are considered: Hawk H, which never withdraws, and Dove D, which withdraws if there are two individuals in the place. Since if both individuals remain in the place they both die, the pay-off of Hawk in the presence of another Hawk equals zero. When two Doves meet, one of them withdraws, so that on the average half of them stays.

	H	D
H	0	1
D		

to withdraw, even if faced with the danger of death due to overcrowding. This result should make us very careful when we postulate any kind of self-regulation. For Dove strategy to be an ESS, some additional possibilities of surviving and leaving progeny for individuals migrating outside the grain should be assumed. This assumption is made in section 8.1, where the selective advantage of emigration is discussed.

Look once again at the pay-off matrix in Table 7.1 with the additional assumption that two individuals that meet each other in an indivisible piece of a resource are genetically identical members of the same clone. If the strategy is genetically determined, then Hawk always meets Hawk, while Dove meets Dove. To see which strategy is an ESS, we have to compare the pay-off for Hawk in the presence of another Hawk $E(H, H)$ with the pay-off for Dove in the presence of another Dove $E(D, D)$.

This comparison clearly shows that for the members of the same clone, D is an ESS. This result is very robust. If we assume that the death of two Hawks is not inevitable, so that each of them may survive with probability $s < 1/2$, D is still an ESS. Let us now make another assumption, that the groups of Hawks and Doves are not genetically homogeneous, so that a Hawk may appear among Doves, or vice versa, because of mutation or immigration, with probability P. With this assumption, the average pay-off for Hawks equals P, whereas the average pay-off for Doves equals $(1-P)/2$, which implies that D is an ESS if $P < 1/3$.

Competition among two individuals belonging to the same clone represents a very special case, but in nature this is not rare. It is generally accepted that competition among genetically identical individuals is the strongest kind, because these individuals are expected to have identical ecological requirements. But there is another side to genetical identity. If individuals are genetically identical by descent, like members of the same clone, we may expect kin selection to act toward withdrawal under strong competition for resources. On the other hand, I do not know of any good empirical data confirming the theoretical conclusions concerning competition among the members of the same clone, as presented above. Some related data on migration of *Hydra* polyps are presented in section 9.3.

It is much more complicated to analyze the existential game presented in Table 7.1 for the intermediate case of two full sibs (Łomnicki 1980a), which represents a game between relatives (Maynard Smith 1982). Assume first that recessive homozygotes *aa* adopt strategy D, while heterozygotes *Aa* and dominant homozygotes *AA* adopt strategy H. Assuming random mating and random formation of pairs of contestants, the proportion of all possible pairs of genotypes made of full sibs can be determined from the frequency q of recessive allele a, as shown in Table 8.2 in Chapter 8. After including the survival of each individual according to the pay-off matrix (Table 7.1), the frequency q_1 of allele a in the next generation is given by the following equation:

$$q_1 = \frac{3(1-q)^2 + 8(1-q)q + 4q^2}{7(1-q)^2 + 12(1-q)q + 4q^2}$$

From this equation, $q_1 > q$ if $0 < q < 1$, which implies that D is an ESS. If we assume that recessive homozygotes aa adopt strategy H, and that the remaining genotypes AA and Aa adopt strategy D, then the final result is altered, yielding stable genetic polymorphism with the equilibrium frequency of allele A, $p = 0.59$. This is presented in more detail in section 8.1.

This latter complication, due to the assumption that D strategy represents a dominant genetic character, does not change the main conclusion derived from the analysis of the game between two full sibs. The ESSs for this game are clearly different from the ESS for unrelated, randomly chosen individuals. If there is a shortage of resources, the withdrawal of some individuals does not have to be a parental manipulation, but may be an adaptation evolved by kin selection. The models that assume that by giving up its place an individual is gaining something else in a different place are discussed in Chapter 8. Here it was assumed that an individual gains nothing and loses everything, and as shown above, such behavior can evolve only among closely related individuals.

The difference among the three models discussed above can be represented in the form of diagrams (Figure 7.1) that depict the proportion of three different pairs made of Doves D and Hawks H when pairs are sampled randomly (R) from a large population, when they are made of full sibs (S), or when they belong to the same clone (C). According to the existential game presented here, the proportion of Doves D in the next generation depends on the genetic relationship between individuals within pairs. The diagrams in Figure 7.1 can be regarded as graphic representations of self-regulation within trait-groups, as defined by Wilson (1980). These diagrams apply only to trait-groups

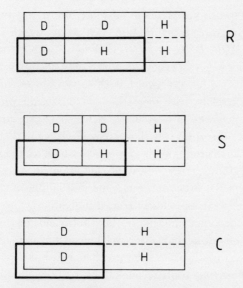

FIGURE 7.1. Three diagrams representing different proportions (horizontal axes) of pairs of individuals that represent two genotypes, D and H. These pairs can be randomly formed according to binomial distribution (R), be made of full sibs (S), or represent members of the same clone (C). For these diagrams it was assumed that half of the population represents genotype D, the remaining half genotype H. The rectangle at the bottom left-hand corner of each diagram includes the individuals that survive, accroding to the model presented here.

of two individuals, and for a given frequency of genotypes adopting the strategies D and H. Therefore no general conclusion can be derived from them. Nevertheless, they enable us to see more intuitively the mechanism by which trait-group selection may promote population self-regulation.

Note that the survival of D strategists, as assumed in this model and shown on the three diagrams (Figure 7.1), is due to the existence of homogeneous groups in which both individuals are D strategists. The higher the variance among groups, the smaller it is within groups, and the larger the proportion of pairs made up of individuals of the same genotype. As Wilson has shown, a decrease in variance among trait-groups diminishes the effectiveness of trait-group selection. Diagram (R) of Figure

7.1 can help us to appreciate this more intuitively. Instead of two, imagine three individuals in each group, which will then be *DDD*, *DDH*, *DHH*, or *HHH*. If *D* occurs with frequency Q, then the frequency of groups in which only *D* strategists occur equals Q^2 for groups of two, but Q^3 for groups of three individuals. Therefore, with increasing group size, the proportion of groups made exclusively of *D*, in which *D* can increase in frequency, decreases. This shows that if self-regulation cannot evolve for small groups made of two individuals, it cannot evolve for larger groups either. Thus self-regulation cannot evolve in groups formed at random, but some possibilities of self-regulation do exist among closely related individuals.

For the theoretical models presented above, no good empirical confirmations are available. It is well known that animals exhibit a lower level of aggression toward their kin, but there are no data to show that contest competition can be a result of kin selection.

7.2. SELF-REGULATION IN CONFINED LABORATORY POPULATIONS

Old textbooks of ecology, for example Allee et al. (1950), extensively discussed population growth as a function of density (i.e. the number of individuals per unit area or per unit volume) and often disregarded limiting factors other than space. In the ecological literature, there are plenty of data concerning population growth in limited space. This chapter discusses only those in which space is the *only* limiting factor, while all other resources that are required for survival and reproduction are in excess. To make the arguments clear, I will put aside all data on density in relation to available food, nesting sites, concentration of oxygen or carbon dioxide in water, as well as data on the density of sedentary organisms, for which an area of available substratum space can be a real limiting factor.

A separate problem is that population density can manifest itself by the concentration of metabolic products or other chemical substances that are secreted by organisms into water, soil, or other media in which they live. This is so-called water conditioning (or as in the case of *Tribolium* populations, flour conditioning), which may exert either positive or negative effects on population growth. Putting aside the positive effect of conditioning, the negative effect can be interpreted in the following two ways.

One can imagine that a certain level of concentration of metabolic products makes the normal functioning of an organism impossible and decreases both natality and the probability of survival, for strictly physical reasons, in the same way that the shortage of oxygen hinders respiration and the shading of light hinders photosynthesis. I am using here the word "physical," not "physiological," because physiological reasons might be adaptations evolved by natural selection, not physical necessities. If so, then conditioning is an ultimate limiting factor for a given organism. On the other hand, it is possible that conditioning is in fact neutral for reproduction and survival, but that it acts as a signal to inform organisms about their population density. If so, conditioning ought to be treated as a proximate factor, and the reaction to a certain level of conditioning should be regarded as an adaptation that allows organisms to adopt the best strategy at high population density. Thus, under certain circumstances, they may react to such a signal, but under others they may ignore the same signal.

It is sometimes very difficult to decide which of these two interpretations is the correct one. If the reaction to conditioning is physiological, one can hardly tell whether this is a physical necessity or an adaptation. Besides, conditioning means the presence of a mixture of various chemical substances, some of which are very unstable. If we intend to study the reaction of organisms to their density, then the behavior of individuals at various densities seems to be a much better object of investigation

133

than medium conditioning. Behavioral reactions can be identified much more readily, and it is possible to decide easily whether we are dealing with proximate or ultimate factors.

Studies of confined populations of small rodents—mice, rats, and those exhibiting periodic population cycles in nature—are especially numerous. The most common design of these experiments is as follows: A small number of individuals of both sexes is placed in a cage or a room, with food and water supplied *ad libitum*, and then over tens or hundreds of weeks the number of individuals, their sex, and the number of newborn and dead are recorded. Very often animals are marked individually and their physiological status is monitored.

In such confined experimental populations, there is at first an increase in the number of individuals. Individuals react to increasing population density with decreased natality, and sometimes poor survival, which leads to the arrest of population growth, much below the level determined by food supply. At this time, population size can be very stable, as in the case of prairie deermice *Peromyscus maniculatus*, in which the population consisted of 13 individuals for more than 300 days (Terman 1980). Such populations can also exhibit oscillations in size, as reported by Petrusewicz (1957) for laboratory populations of house mice. In the latter, after a period of growth, a drop in population size was reported, followed by irregular oscillations and periods of stable population size lasting for several weeks. In both cases, there was no shortage of food, water, or nesting sites; therefore it can be concluded that density itself was the factor that regulated population size. It is important to note that in spite of identical experimental conditions, the equilibrium densities in different experimental populations do differ; for example, Terman obtained densities ranging from 6 to 47 individuals in cages of 1.82 square meters each.

There are two different approaches to explain the experimental results described above: the holistic approach and the reductionistic approach. If one takes for granted that many various interrelations exist among the members of a population, so that a population manifests itself as a structure too complex to be understood, then one has to regard a population as an integrated entity or as a black box. One can study external reactions of this entity to different external conditions, but for the present, one is unable to understand how it works inside. This approach seems irrational today, but it was not irrational for those who treated a population as an entity similar to a single organism, in the way discussed in section 1.1. It was said that a population is an entity able to limit its growth in a manner similar to that of an individual organism, and that population structure determines the limit of growth. Very often it was argued that the ability of a population to limit its growth before the shortage of resources occurred was obvious evidence for self-regulation, the function of which is the avoidance of resource depletion.

Reductionists take a different approach. They try to find out how individuals perceive the presence of other individuals, and how this perception modifies their behavior and physiology, which in turn determine their reproduction and survival, and consequently their population density. Cohen et al. (1980), summarizing a series of papers on self-regulation among mammals, listed dozens of different signals of density to which different species under various conditions react in dozens of ways. The results of these behavioral and physiological reactions are decreased natality and increased mortality of young individuals, which eventually lead to population self-regulation. This self-regulation can be very precise (although this is not always so, since populations sometimes oscillate in size); sizes of self-regulated populations may differ, even if they are kept under almost identical experimental conditions; and self-regulation occurs much below the level of the shortage of resources, such as food, water, and nesting sites.

The studies of behavioral and physiological mechanisms that lead to self-regulation of density in laboratory populations of small rodents are interesting and important. For a full understanding of this process, however, we need something more than the knowledge of mechanisms; for example, there is the important question of how such mechanisms could have evolved and how they are maintained by natural selection. Why do mature females refrain from reproduction? Why are embryos resorbed? Why are newborn animals cannibalized by their parents? If we reject the possibility of adaptation that evolved by group selection in order to avoid population extinction following resource depletion, we have to find some other answers to these questions.

Density is not an ultimate limiting factor for small rodents, but rather a proximate one. There is obviously no shortage of air, or any other resource, in laboratory populations of small rodents, and therefore the behavioral and physiological reactions to overcrowding exhibited by individuals in these populations are adaptations. How were these adaptations brought about? An attempt to answer this question is given below.

7.3. OPTIMAL REPRODUCTION IN POPULATIONS WITH UNEQUAL RESOURCE PARTITIONING

Small rodents are able to reproduce many times during their lifetime and a few times within one season. Their populations are often subject to violent oscillations in size. Applying the theory of life history strategies, the optimal age of first reproduction, the frequencies of litters, and the size of those litters can be inferred on the basis of age-specific survival rates and the costs of reproduction. However, within the framework of this theory, no theoretical models have been developed to account for two other phenomena: unpredictable changes in population density in relation to available resources, and unequal resource partitioning.

In order to show how the changes in density and unequal resource partitioning may influence reproduction, I will present here a simplified graphic model in discrete units of time (Figure 7.2). For this model the following assumptions are made: (1)

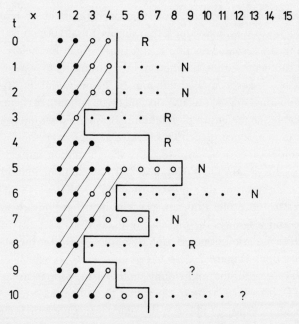

FIGURE 7.2. Changes in the size of a hypothetical population with a randomly varying food supply (solid line) and hierarchy among individuals expressed by their ranks x, during ten time moments t. Solid circles denote individuals that survive and reproduce giving two offspring; open circles denote those which survive but do not reproduce; dots denote individuals that are unable to survive to the next time moment due to a shortage of resources. The letter R denotes moments of time in which reproduction is of selective advantage, the letter N those in which reproductive effort is wasted. A question mark stands at those moments of time for which it cannot be decided whether reproduction is advantageous or not.

An individual born at time t is able to reproduce at time $(t + 1)$. (2) The amount of resources available for the population changes randomly, in such a way that it can support $k = 2$, 4, 6, or 8 individuals. (3) If an amount of resources can support k individuals, then $k/2$ individuals are able to reproduce and to survive to the next moment of time, while the remaining $k/2$ individuals are unable to reproduce but can survive to the next moment of time. (4) Unequal resource partitioning is manifested by linear social hierarchy, with reproducing individuals having the highest ranks, individuals that survive but do not reproduce having the second best ranks, and those which die before the next moment of time having the lowest ranks. This linear hierarchy persists from one time unit to the next, and the fate of an individual depends on its rank and on the amount of resources available for the entire population. (5) Each individual that is able to reproduce gives two offspring, which take the lowest ranks in the hierarchy. (6) Mortality of high-ranking individuals is introduced into the model in such a way that one individual of the highest rank dies in every time unit, which allows all the others to move one rank up. (7) The model includes females only, and it ignores the presence of males.

The simulation of population dynamics, based on the above assumptions (Figure 7.2), starts at time $t = 0$, at which there are resources for four individuals, and four individuals are present. Two of them reproduce, giving two offspring each, and these four offspring appear at the time $t = 1$. One individual of the highest rank dies, and therefore there are seven individuals at time $t = 1$. Since at $t = 1$, the amount of resources is sufficient for four individuals only, the three lowest-ranking individuals die. Analogically we may simulate the future dynamics of this population, as determined by the amount of available resources and the assumptions listed above. If there is a sudden increase in the amount of resources, as between moments $t = 3$ and $t = 4$, then the assumed reproduction is not sufficient to increase population density to the maximum level possible. If there is a sudden decrease in the amount of resources, then more individuals with lower ranks die.

The ability of newly born offspring to survive at different time moments is the important property of this model. Note that individuals born at times $t = 2$, 5, and 7 die in the next time unit, whereas individuals born at times $t = 1$ and 6 survive to the next time unit but eventually die without leaving progeny. If there is any cost of reproduction, then the reproduction at the moments mentioned above is a complete waste of parental effort. Reproduction makes sense if progeny can survive and give birth to their own progeny, as in times $t = 0$, 3, 4, and 8. This is so because after these time moments either there is an increase of the amount of resources or at least there is no decrease of their amount. Therefore, the organism's decision to reproduce or to refrain from reproduction should be based on the perception of the number of other individuals in the population in relation to available resources.

This simple graphic model cannot serve as a precise explanation of the phenomenon of limited reproduction. Ten random numbers were used here to simulate random changes in the amount of resources, but with other random numbers the result may be slightly different. Nevertheless, if the amount of resources is a random variable with a constant mean value, then after a very good period with resource abundance, a decrease of the amount of resources is to be expected. What is even more important, the amount of resources is not really a variable independent of population size, and we may expect that after high population density there will be a reduction in the amount of food available. Therefore high population density is a signal of poor prospects for newly born individuals. If the amount of resources changes in an unpredictable way, then population density is the only reliable signal of future per capita resource availability.

Population densities during the experiments on population self-regulation described in section 7.2 were many times higher than the densities that occur in the field, and therefore they were signals of poor prospects for newly born animals. Considering that the cost of reproduction and care of young may considerably diminish life expectancies and future reproduction, it seems to be of selective advantage to refrain from reproduction or from taking care of offspring at times of poor prospects for the survival of these offspring. If so, then refraining from reproduction, foetus resorption, and cannibalism of progeny can be regarded as adaptations that promote higher reproductive output at more favorable times.

If one accepts the above interpretation, the phenomenon of self-regulation in confined laboratory populations of small rodents might be regarded a by-product of the existence of adaptations such as those described above. At high population density, refraining from reproduction and from copulatory behavior can result in a decrease of the weight of the reproductive organs of both sexes. For example, according to Terman (1980), in laboratory populations of prairie deermice *Peromyscus maniculatus bairdi*, which are subject to self-regulation, the mean weight of testes equals 69.98 mg, compared with 202.24 mg for the control group, while mean weight of uteri equals 6.09 mg, compared with 25.14 for the control group. These large and statistically significant differences were not caused by the poor condition of animals that refrain from reproduction; differences in body weight among these groups, among both males and females, are small and statistically insignificant. This supports the hypothesis that for these mice, refraining from reproduction was not a necessity due to the shortage of resources, but an adaptation to predicted future poor conditions.

The fact that we are able to select strains of mice that are tolerant to the presence of their conspecifics is another confirmation of the hypothesis that behavioral reactions leading to self-regulation are adaptations, not necessities. For example, Lloyd (1980) studied two laboratory populations of prairie deermice: the first population was derived from stock kept for 30 years in the laboratory, the second was composed of individuals caught recently in the field. Their population sizes at the state of equilibrium were 87 and 15 individuals, respectively. This shows that intolerance toward other conspecifics, on which the self-regulatory behavior is based, can be altered by selection and must be regarded as an adaptation.

Self-regulation in laboratory populations of small rodents is related to many behavioral and physiological phenomena that are not included into the simple graphic model given by Figure 7.2. Nevertheless, this model allows for the prediction that self-regulation is possible if the two following conditions are satisfied:

(1) a self-regulating population should consist of iteroparous organisms, for which refraining from reproduction at present can be compensated by a higher reproductive output in the future, and (2) there must be unequal resource partitioning expressed by social hierarchy in which the youngest individuals have the lowest ranks. Without a social hierarchy, young individuals always have the chance to survive, and it is never advantageous to refrain from reproduction.

The first condition is fulfilled for rodents, which are iteroparous; self-regulation has not been reported to occur among semelparous organisms. The second condition is also known to be fulfilled among laboratory populations of small rodents. Just after the establishment of a new population, there are frequent fights among males; then, after a lapse of time, the fights cease and a social hierarchy is established. If social hierarchy is a necessary condition for self-regulation, then the disturbance of such a hierarchy should lead to population growth. This result was obtained by Petrusewicz (1957, 1963) in laboratory populations of house mice. These populations exhibited oscillations in their sizes with spontaneous rapid population growth and a relatively slower decline in number, but a population increase could be provoked by the disturbance of its social structure. This was done either by changing the cage (Petrusewicz 1957) or by introducing four to seven virgin females into the studied population, for a period of one week (Petrusewicz 1963). In both cases, such disturbances caused statistically more frequent occurrences of population growth. The cages were changed for others of the same size, larger, or smaller—even 11 times smaller—but in all cases population growth occurred more frequently than in control populations. This is an interesting result, because it shows that the disturbance of social structure can bring about population growth, even if the space becomes smaller. Artificially induced population growth is usually initiated at a lower density than spontaneous growth, and it is caused by both higher natality and better survival of sucklings in nests. The first reaction to a disturbance is more aggressive behavior, which sometimes leads to a small population decline, after which growth occurs. Petrusewicz (1963) found that growth frequently occurs after the death of a single adult individual.

The phenomena described above can be interpreted applying the diagram in Figure 7.2. Imagine that an individual of the highest rank $x = 1$ does not die in every time unit, as proposed in the model, so that new places are not available in some time units. Under such circumstances, the only chance for the survival of newborn individuals is an increase in the amount of resources. Under normal field conditions, a density as high as that used in the experiments is a signal of food shortage in the nearest future; therefore, to refrain from reproduction seems to be an optimal strategy. The strategy changes with the death of a dominant individual or with any disturbance that gives some prospects for the survival of younger individuals. This explains why females then start to reproduce and to take care of their offspring, which leads to population growth.

The phenomenon of self-regulation in laboratory populations of small rodents is an interesting one, but it only occurs under special laboratory conditions. This phenomenon is nothing but a by-product of adaptations by which individuals can maximize their lifetime reproductive success by refraining from repro- duction under overcrowding. Density is a proximate factor here, a signal of some ultimate factors, such as a shortage of food in the nearest future. In laboratory populations, such an ultimate limiting factor does not exist, and therefore animals kept for many generations in the laboratory evolve tolerance to high density. The important conclusion to be drawn from these stud- ies is that population growth may sometimes be brought about by a disturbance of behavioral relations between individuals, and does not necessarily result from an improvement in eco- logical conditions.

CHAPTER EIGHT

Emigration and Unequal Resource Partitioning

As suggested in Chapter 6, contest competition among animals is possible if some individuals withdraw, giving way to others. If such withdrawals are not physical necessities but adaptations, the question arises: What are their selective advantages? As reviewed in section 7.1, withdrawals that provide no gains at other times or in other places cannot evolve in groups of unrelated individuals, but they can arise in groups of close relatives. Withdrawal can also take the form of an escape in time—as, for example, refraining from reproduction, which is discussed in section 7.3—or an escape in space. Can an escape in space (i.e., emigration) be a mechanism of population regulation? I will discuss this question in the present chapter.

8.1. EMIGRATION FROM GROUPS OF RELATED AND UNRELATED INDIVIDUALS

In section 7.1, the model of the conflict between two individuals was discussed, in which it was assumed that an individual that has given way dies without leaving progeny. Here an extended version of this model is considered, which assumes the possibility that a withdrawing individual will survive with probability s; hence if two Doves meet in one place, the pay-off of the one that stays equals unity, while that of the one which emigrates equals s. On average, if Dove meets Dove, its pay-off equals $(1 + s)/2$. This version may be regarded as a model of withdrawal to a worse place within the same local habitat or as a model of emigration outside such a habitat.

The pay-off matrix of this version (Table 8.1) shows that neither pure Hawk strategy H (staying in the place in the presence of another individual) nor pure Dove strategy D (withdrawing if there is another individual) is an evolutionarily stable strategy (ESS). This is because a population of Hawks can be invaded by a Dove mutant, or a population of Doves can be invaded by a Hawk mutant. The fitness of the strategies considered depends on the proportion in which these strategies are adopted. The ESS for this pay-off matrix is a mixed strategy, the ESS being to adopt strategy D with probability P as given by equation

$$P = 2s/(1+s), \qquad (8.1)$$

where s denotes the probability of survival of the individual that withdraws. The value of P can be interpreted as both the ESS

TABLE 8.1. Pay-off matrix of the game between two individuals in a place that can support one individual only, with a lower probability of survival for the individual that withdraws. Two strategies are presented here: Hawk H, which never gives up, and Dove D, which withdraws in the presence of another. Note that the probability of survival of the individual that withdraws equals $s > 0$, whereas for the game presented in Table 7.1, it equals zero.

	H	D
H	0	1
D	s	$(1+s)/2$

probability of adopting the Dove strategy or the ESS proportion of Dove strategists within the population.

Equation (8.1) shows that if $s = 0$, this model is reduced to the model presented in section 7.1, whereas if $s = 1$, each individual that is not solitary should emigrate. For intermediate values of s, P is slightly higher than s (Figure 8.1). If the probability s of migrant survival is very low, then according to the prediction of this model, the decision to withdraw from a place in which there are enough resources for only one individual is very unlikely.

As was shown in section 7.1, even when an emigrant has no possibility of survival, withdrawal may be an adaption if the rivalry is among full sibs and if the strategy of Dove D is genetically determined by recessive homozygotes. If Dove strategy is determined by dominant homozygotes AA and heterozygotes

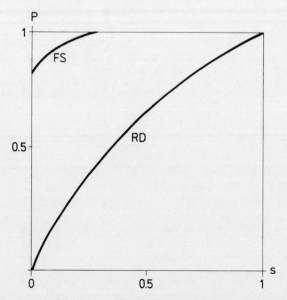

FIGURE 8.1. The ESS proportion P of individuals that withdraw in the presence of another, for pairs formed randomly (RD) and for pairs made of full sibs (FS), as the function of the probability s of migrant survival.

145

Aa, and emigrants survive with probability $s > 0$, then the model of the game between full sibs becomes more complicated. If we assume that both mating and the formation of pairs among full sibs are random, the proportions of each type of pairs can be calculated and the survival of each individual of every pair type determined, as shown in Table 8.2. From these proportions, the frequency p of allele A responsible for Dove strategy can be calculated for the next generation. The frequency p increases if the inequality

$$p^2(1 + s) - 4p(1 + s) + 2 + 6s > 0 \tag{8.2}$$

is fulfilled. Thus, the frequency p_e of the allele A at equilibrium is given by the following equation:

$$p_e = 2 - \sqrt{2(1 - s)/(1 + s)}. \tag{8.3}$$

TABLE 8.2. The frequencies of the groups of two individuals that are full sibs, and the fate of each genotype, when assuming random mating and random formation of pairs within families. Dominant homozygotes AA and heterozygotes Aa are assumed to be emigratory genotypes; recessive homozygotes aa are assumed to be nonemigratory. If both individuals in a group have emigratory genotypes, then in half of the cases one genotype emigrates, while in the other half the second genotype emigrates. Therefore, to simplify the presentation, the groups that are made of two different genotypes are listed twice. The following fates of each individual are considered: survival in the place (1), survival as an emigrant (s), and death in the place if neither of the two individuals emigrates (0).

		1	s	0
AA, AA	$p^2(p + q/2)^2$	AA	AA	—
AA, Aa	$p^2q(p + q/2)$	AA	Aa	—
AA, aa	$p^2q^2/4$	aa	AA	—
Aa, AA	$p^2q(p + q/2)$	Aa	AA	—
Aa, Aa	$pq(p^2 + 3pq + q^2)$	Aa	Aa	—
Aa, aa	$pq^2(p/2 + q)$	aa	Aa	—
aa, AA	$p^2q^2/4$	aa	AA	—
aa, Aa	$pq^2(p/2 + q)$	aa	Aa	—
aa, aa	$q^2(p/2 + q)^2$	—	—	aa, aa

If $s = 0$, then equation (8.3) describes the situation discussed in section 7.1; if $s \geq 1/3$, then $p = 1$, which implies the fixation of allele A. This latter case means that an individual should always emigrate from a place that can support one individual only, thus leaving his full sib, even if the probability of migrant survival is three times lower than that of a nonmigrant.

The difference between randomly formed pairs and pairs made of full sibs in the ESS proportion of emigratory genotypes is shown in Figure 8.1. For randomly formed pairs, this proportion is the ESS proportion of Doves, or the probability that an individual will adopt Dove strategy, while for pairs made of full sibs this is the proportion of both AA and Aa genotypes representing the Dove strategy. If Dove strategy is determined by recessive homozygotes aa, then their allele a is fixed and the proportion of Dove strategists equals unity for all values of s, as shown in section 7.1. The Dove strategy can be regarded as a strategy of emigrants that leave a place in which there are resources for one individual only or as a strategy of voluntary withdrawal into a worse position in the same place, in order to prevent the depletion of resources and the death of both individuals. Figure 8.1 clearly shows that the decision of voluntary withdrawal into a worse position depends on the relatedness of the individuals facing the danger of overexploiting the resources available to them.

Hamilton and May (1977) proposed a different model of emigration out of the sib groups. They assumed, not the possibility that the entire group could become extinct, but rather the possibility that only one of the individuals in each group could survive. The problem solved by their model is as follows: Is it of selective advantage for an individual to compete within the group of sibs, or is it more advantageous to take the risk of migrating to another place, to enter a group of individuals unrelated to him and to compete with them? It seems of no advantage for an individual to undertake the risk of migrating to another place if that place is equally crowded and if the arrival of immigrants decreases even further the chance of obtaining this place. On the other hand, if the groups consist of the progeny of parthenogenetically reproducing individuals, which means that they are clones, then the evolutionarily stable proportion v of emigrating individuals is, according to these authors, given by

$$v^* = 1/(2 - s),$$

where s is the probability of migrant survival. If the groups consist of full sibs, that are the progeny of sexually reproducing parents, then the ESS proportion v of emigrating individuals is given by

$$v^* = 0,$$

for $0 < s \leq 0.5$,

$$v^* = (2s - 1)/(4s - 1 - s^2),$$

for $0.5 < s < 1$.

The results shown above indicate that it is of selective advantage to emigrate to avoid competition with close relatives even if prospects for migrants are poor. For parthenogenetically produced progeny, half of the individuals should emigrate, even if the probability of survival during the migratory episode equals zero.

It is a well-known fact that the seeds of many plants and the juvenile forms of many animals exhibit either morphological or behavioral adaptations for dispersal away from the places in which they were born. On the other hand, there are no good experimental data to show that the tendency to emigrate from groups of closely related individuals is much stronger than from unrelated groups. One of the hypotheses explaining population cycles of small rodents, namely that of Charnov and Finerty (1980), is based on the assumption that emigration occurs mainly from groups composed of closely related individuals, and that it is less common from groups of unrelated individuals. According to this hypothesis, at very low population densities the groups consist of closely related individuals that exhibit a high reproductive rate and a high emigratory tendency. After a population attains a certain density at which there are no empty places to colonize, some migrating individuals enter other groups, which then become groups of unrelated individuals. This inhibits emigratory behavior and increases aggressiveness within groups, which in turn results in a reduction of natality and survival, leading to low density. This is an interesting hypothesis, although as yet it has not been presented in the form of a mathematical model, nor has it been experimentally confirmed in the field.

Thus far in this section, only symmetrical games have been considered, in which both opponents or all opponents have the same chance of surviving and leaving progeny. Furthermore, it is assumed here that if two individuals remain in a place in which there are resources for one individual only, both will die. A simultaneous death of both competitors implies that the two are very similar in their strength and their ability to acquire resources. If we relax this assumption, and if we allow for differences between two individuals in one place such that the weaker individual survives with probability $w(0 < w < 0.5)$ and the stronger one with probability $1 - w$, then the pay-off matrix changes to the form shown in Table 8.3. The result of this game is a trivial one; nevertheless, it is important: if the probability of survival of the weaker individual, w, is lower than the probability of migrant survival, s, then the weaker individual should emigrate; otherwise it should stay.

In this game we take for granted that an individual is able to estimate whether it is stronger or weaker than its competitor. This can be achieved during behavioral interactions such as ritual fights, during which individuals may learn what is the optimum choice for them. It is possible that the shoving contest between stem mothers of *Pemphigus betae* described in section 6.3, and other similar types of agonistic behavior, are not real fights but rather behavioral interactions aimed at obtaining information about the opponent's strength.

Table 8.3. Pay-off matrix for the asymmetric game, in which the weaker individual survives in a certain place with probability w $(0 < w < 0.5)$, the stronger one with probability $(1 - w)$. Two strategies are assumed here: Hawk H, which never withdraws, and Dove D, which withdraws if it happens to be the weaker individual. The pay-off to the stronger individual is given in the upper left corner of each entry, the pay-off to the weaker one in the lower right corner. In the center of each entry, the mean of these two values is given, since it was assumed that one of two individuals is stronger and the other weaker. Dove is an ESS if the probability s of migrant survival is higher than the probability w of survival in the place. Otherwise, Hawk is an ESS.

	H	D
H	$(1-w)$ ⟍ $1/2$ ⟍ w	1 ⟍ $(1+w)/2$ ⟍ w
D	$(1-w)$ ⟍ $(1-w+s)/2$ ⟍ s	1 ⟍ $(1+s)/2$ ⟍ s

8.2. IMPERMANENT LOCAL HABITATS
IN HETEROGENEOUS SPACE

Animal migration will be discussed here in terms of a model of many local habitats, surrounded by more or less hostile space. These habitats may or may not be occupied by local populations. A local population may become extinct, either by the disappearance of its local habitat or by extinction of the entire local population, after which its habitat can be colonized again. Assuming the state of equilibrium, a disappearance of a local habitat in one place is balanced by a formation of another local habitat somewhere else. Such a system of local populations, called a metapopulation by Levins (1970), was also applied by other authors in theoretical models of group selection (Maynard Smith 1964; Gilpin 1975).

If local populations are not permanent, then migration is necessary for the persistence of the metapopulation—in other words, for the survival of a species in a given area. This does not necessarily imply that the function of migration is to prevent species extinction, which would require selection on a species level. If we assume that selection among individuals is the most important force of evolution, then when explaining the evolution of migration, mechanisms other than species selection should be invoked.

It is assumed here that in a local habitat, physical and biotic conditions are sufficiently favorable to make population growth possible, so that the number of emigrants is always higher than the number of immigrants. Gill (1977), studying populations of red-spotted newts *Notophthalmus viridescens* inhabiting a set of small mountain ponds, found that reproduction was unsuccessful in most of these ponds. In fact, more than 90 percent of young individuals inhabiting all these ponds had completed their development in the same single pond. This one favorable pond should be regarded as the local habitat, all others as part of the surrounding hostile space.

Defining a measure of persistence of local habitats is an important point. Southwood (1962) has proposed estimating this persistence in relation to the generation time of a species occupying the habitat. In an extreme case, animals must change local habitats several times during their lifetime. This is related to the spatial scale: a single plant or a meadow on which these plants occur may be regarded as local habitats for herbivorous insects. Considering the enormous variation in life histories and ecological requirements of different species, all definitions and classifications of local habitats are more or less arbitrary. For further discussion, I will define a local habitat as a part of space that is permanent enough and large enough that the population of the species in question may persist within its boundaries for at least one season, or for the duration of its entire live stage.

151

If an individual utilizes a place that lasts for a much shorter time, such a place should not be classified as a local habitat, but as a patchily distributed feeding or oviposition site.

There is one important feature of the persistence of local habitats that cannot be described by their mean longevity. For example, a cow pat not only lasts for a relatively short time as a suitable habitat for some fly larvae, but its duration is age-dependent, and therefore more or less predictable. In a similar way we may estimate how long other dead fragments of plants and animals will last. Animals may be able to perceive some environmental signals indicating how long a habitat can support them, and thus be able to leave it before its complete deterioration or the depletion of their resources. This case is quite different from that of a water pool in a forest, which may dry out within a week or may just as well last for several years. Imagine a meadow, surrounded by a forest, that is a local habitat for a butterfly species. An accidental frost or a storm could kill all individuals of this species in the meadow, so that for a very short time this particular place would cease to be the local habitat for this species. The result of such a catastrophe for the local population is the same as if this habitat became extinct. On the other hand, if no adverse accidents occur, this local habitat may last for years. The impermanence of the habitat on this meadow is not due to the limited amount of the resources, but to the inhabiting species' susceptibility to adverse physical conditions.

It can be said that the cow pat is a local habitat with determinate duration, whereas the meadow is one with an indeterminate duration. More precisely, the cow pat, or other dead matter that constitutes a local habitat, exhibits an age-dependent duration, whereas the survival of a habitat with an indeterminate duration is age-independent and is a negative exponential function of time.

We may expect different kinds of individuals to emigrate from habitats of determinate duration than from those of indeterminate duration. If the duration of the habitat is determinate, and if only one life stage, for example larvae, lives in it, then the strongest individuals that are able to complete their development before the resources are depleted should be the first to emigrate. In a habitat of indeterminate duration, which can support an animal species during its entire life cycle, there is no reason for the strongest individuals of this species to emigrate. The strongest individuals should also refrain from emigrating from a habitat of determinate duration if they are far from completing their development.

Emigrating from habitats of determinate duration before resource exhaustion is a necessity, and it cannot work as a mechanism of population regulation. The theoretical concepts of population regulation by emigratory behavior (Lidicker 1962; Wynne-Edwards 1962) refer to emigration from habitats with indeterminate duration. On the other hand, most laboratory experiments, like those carried out on flour beetles or fruit flies, are concerned with species that in nature live in habitats with determinate duration. Only under some special laboratory arrangements, with a permanent inflow of new resources, can the habitats of these species be considered habitats of indeterminate duration.

We can imagine three different models of emigration from local habitats:

(1) *Emigration into the space outside local habitats, where mortality is not higher than within them.* In this case emigration cannot act as a mechanism of population regulation, but is a mechanism that brings about more uniform distribution of individuals in relation to available resources. The formal description of this type of migration is provided by the theory of ideal free distribution, discussed in section 8.6.

(2) *Emigration of a surplus of individuals.* Considering the differences among individuals in their ability to acquire resources, we can easily imagine that there are individuals that have no possibility of surviving and reproducing in a given local habitat. If their chances of survival in the local habitat are zero, then it will be of selective advantage for them to emigrate, even if

the probability of surviving the migration episode is very low. If some individuals have no chance of surviving in their local habitat, then for the dynamics of their local population it is irrelevant whether they die in the local habitat or leave it. When this model is accepted, emigration is of no importance for population dynamics, and the evolution of migratory behavior is a trivial problem. On the other hand, experiments on migration from experimental populations of some species (Chapter 9) clearly show that emigration does influence population dynamics and results in a decline of population density as compared to confined populations. Therefore the emigration of the surplus of individuals, as described here, does not seem to have a counterpart in nature.

(3) *Emigration and immigration regulating population density.* If emigration is to act as a mechanism able to regulate density below the level of a confined population, two conditions must be simultaneously fulfilled. First, emigration should take place when there are still some resources available. If for some reason emigration is impossible, population density should be appreciably higher than if emigration had occurred. Second, the space between local habitats must be very hostile, considerably reducing the number of migrants before they are able to colonize new habitats. A reduction need not necessarily occur in the space between local habitats; it could also result from difficulties in entering already occupied habitats. This implies that regulation of population density can be brought about not only by density-dependent emigration, but also by density-dependent immigration. Population regulation by immigration was postulated, for example, by Krebs and Perrins (1978) for the great tit, *Parus major*. This requires that immigrants be unable to enter an already fully occupied habitat and need to seek an empty or partially empty one.

Emigration that occurs before resources are exhausted, as described here, can be defined after Lidicker (1975) as presaturation migration. The level of density saturation in relation to available resources is determined for the moment in which migration takes place, and not for some future time in which conditions may deteriorate. An example of a presaturation migration by which a food shortage is avoided during the following winter was found by Grant (1978) in the vole *Microtus pennsylvanicus*. Presaturation migration may occur to avoid not only forthcoming seasonal food shortage but also a forthcoming population increase, which eventually leads to food shortage. When postulating presaturation migration with low migrant survival, selective advantage must be clearly explained. This is discussed in the following sections.

8.3. EVOLUTION OF EMIGRATION FROM LOCAL POPULATIONS WITHOUT UNEQUAL RESOURCE PARTITIONING

Is presaturation emigration possible from local populations in which all individuals are identical? Within a single population the genotypes that promote emigratory behavior should be selected against, unless the frequency of these genotypes increases due to immigration. Considerable immigration is possible if local habitats are subject to very frequent extinctions followed by colonization, or if there are large reductions in population size allowing for the settlement of individuals from outside.

The evolution of emigration from local habitats can be regarded as a problem of group selection. In the theoretical models of group selection proposed by Maynard Smith (1964), Levins (1970), and Gilpin (1975), selection acts mainly to decrease the probability of extinction of the entire population, not to increase the probability of colonization of new local habitats by emigration. Van Valen (1971) was the first to apply the concept of group selection to the evolution of migratory behavior. In his model, the proportion of emigrating individuals was genetically determined and subject to selection. A numerical solution of his model yields genetic polymorphism for different proportions of emigrating individuals, with the mean proportion of emigrants roughly equal to the probability of extinction of local habitats. This seems intuitively clear, since a higher extinction rate worsens the prospects of nondispersing individuals and increases the chances of migrants, which can find a new empty local habitat more easily. For example, according to Van Valen's (1971) results, if a habitat becomes extinct once in five generations, the proportion of migrants should be close to 20 percent.

The model mentioned above can be classified as a model of group selection because it, like other group selection models, assumes the extinction of the entire local population. Roff (1975) presented a model in which the carrying capacity of local habitats and the reproductive rate within the population varied from generation to generation. He was able to simulate numerically the evolution of dispersal ability and the reduction of population size due to dispersal. His model yields dispersal and reduction of population density for a wide range of parameters, but none of the numerical simulations presented in his paper showed both a high proportion of dispersing individuals and their high mortality. Neither of these simulations allow for both dispersion and mortality to be higher than 50 percent.

A model of the evolution of dispersal with an analytical solution was proposed by Comins et al. (1980). It is an extension of the model by Hamilton and May (1977), mentioned in section 8.1, to groups composed of unrelated individuals. This model determines the evolutionarily stable proportion v^* of emigrating individuals as a function of the population size k of parental individuals that reproduce in the local population, the total number N of their progeny, the probability s of migrant survival, and the probability E of extinction of local populations. The important conclusion of this model is that the proportion v^* of dispersing individuals is a decreasing function of the size k of the parental population. This is because with increasing k, there is a decrease in interpopulation and an increase in intrapopulation genetic variation—that is, the individuals within a single large population are less likely to be related. The number N of the progeny of these k individuals also alters the proportion of emigrating individuals, since with a low reproductive rate there are not enough migrants to colonize new empty habitats, so that those colonizing a new habitat need more than one generation to fill it. This allows for higher reproductive success of the migrants. This conclusion implies that the models that postulate emigration from a small group, for example a group of two individuals as discussed in section 8.1, cannot be generalized to large local populations.

If the size of local populations increases to infinity, then the ESS fraction v^* of emigrating individuals according to Comins et al. (1980) is given by

$$v^* = E/[1 - (1 - E)s], \qquad (8.4)$$

where E denotes both the probability of extinction of a local population and, since the number of all local habitats is assumed constant, the probability of formation of a new empty habitat, while s is the probability of migrant survival during the episode of migration. From this equation, the conditions necessary for the evolution of emigration are population extinctions and formations of new empty habitats. If migrant survival s approaches zero, then the fraction v^* of migrants equals the probability of extinction. With migrant survival s approaching unity, emigration is of selective advantage even if the probability E of extinction is very low. These results were obtained assuming equality among individuals within local populations and competition among them such that an immigrant arriving in an already occupied local population has the same chance of becoming a member of the parental population of k individuals as any other nonmigrating individual in this local population.

An important feature of the model by Comins et al. is that the probability of colonization of an empty habitat is not a linear function of the number of immigrants, as in most theoretical models of the metapopulation, but rather is derived from the Poisson distribution. This feature makes the model mathematically complicated, but also much more realistic than other analytical models of migration. The complexity of the model makes it impossible to introduce additional assumptions, such as unequal resource partitioning among the members of local populations.

The effect of unequal resource partitioning on emigration will be considered in section 8.4, applying a different, much simpler model. A version of this simple model without explicit inequality in resource partitioning is presented below. A qualification should be made here about the resource partitioning among individuals in the models presented in this section. Unequal resource partitioning is assumed neither in the model by Comins et al. (1980), nor in the following one, so that each individual has an identical chance of entering the local parental population of k individuals. However, if there are resources for k parents, it is assumed that only k individuals become parents, irrespective of the number N of competitors for these k places. This already implies inequality and contest competition, because the population size N of young individuals does not affect the population size k of reproducing individuals.

In the following model and in its version presented in section 8.4, two assumptions are made: (1) immigrants can enter only habitats that are empty or unsaturated, so that they do not compete on equal terms with members of local populations, and (2) there are enough immigrants to colonize and fill all empty or unsaturated habitats during one generation.

If in a single local habitat there are sufficient resources for k parents, which reproduce giving R offspring each, then after reproduction there are Rk individuals competing for k places, and without migration the survival of offspring equals $1/R$. With the emigration of the fraction v of individuals, this survival increases to $1/[R(1-v)]$. If a local population becomes extinct with probability E and a new empty habitat is formed with the same probability, then for each newly established empty habitat there are $1/E$ local habitats, both empty and occupied, and $(1/E-1) = (1-E)/E$ occupied local habitats—local populations. This implies that the number of migrants per newly formed empty habitat equals $Rkv(1-E)/E$ individuals, and the probability that an emigrant will find a place to reproduce in an empty habitat equals $k/[Rkv(1-E)/E] = E/[Rv(1-E)]$. It is worthwhile to note that this model does not require the complete extinction of local populations. For example, if there is room for $k = 100$ individuals in a habitat, and the reproductive rate $R = 6$, then after the death of 90 percent of the parents, the total number of progeny $N = 60$, and there is still room for 40 immigrants to enter this population.

To find the evolutionarily stable fraction v^* of emigrants from local populations, an extended ESS model is applied, called "playing the field" by Maynard Smith (1982). The fitness W of a mutant for which the fraction of migrants equals v is given by the equation

$$W = (1-v)/[R(1-v^*)] + vE/[Rv^*(1-E)], \qquad (8.5)$$

in which the first term on the right-hand side denotes the fitness of the fraction of individuals that do not migrate, the second, of those which do migrate. The evolutionarily stable fraction v^* is calculated from condition

$$[\partial W(v,v^*)/\partial v]_{v=v^*} = 0,$$

which after substituting equation (8.5) yields

$$1/[R(1-v^*)] = E/[Rv^*(1-E)]. \qquad (8.6)$$

Intuitively, equation (8.5) describes the fitness of a parent that manipulates its progeny in such a way that v individuals emigrate while the remainder $(1-v)$ stays in a local population. Equation (8.6) allows us to see that the decision to stay or to emigrate can equally well be treated as the decision of an individual. The terms on the left and right-hand sides of equation (8.6) are the probabilities of becoming a member of the parental population, as determined above, for nonemigrating and emigrating individuals, respectively. If the left-hand side of this equation is greater than the right-hand side, then individuals emigrating less frequently reproduce better, decreasing the value of the left side of the equation until the point at which these sides are equal. The same holds true for the opposite case of more frequent migrations. The evolutionarily stable value of fraction v^* is such that it makes the fitness of an individual that stays (left side of this equation) equal to the fitness of one that migrates (right side).

From equation (8.6), the evolutionarily stable fraction of migrants is given by

$$v^* = E, \qquad (8.7)$$

which shows that this fraction is determined by the fraction of populations that become extinct and are substituted by newly arisen empty habitats.

An interesting feature of this model is that the ESS fraction of migrating individuals does not depend on the probability of surviving the migration episode, s. This results from the two assumptions mentioned above. If there are more immigrants than available places, then it does not matter whether an immigrant dies during migration, thus diminishing the number of competitors for empty places, or whether it dies during competition for these places. Formally this model is identical to the model by Comins et al. (1980) given by equation (8.4) in the case of migrant survival s approaching zero.

The result given by equation (8.7) is in agreement with numerical solutions obtained by Van Valen (1971), in which the average migration rate roughly equals the probability extinction of local habitats. The result of the model presented here seems to be more general, since it does not require the action of group selection—in other words, it does not require the complete extinction of the entire local population. Partial extinction allowing immigrants to enter a local habitat can also promote emigration.

This model fulfills one condition that makes population regulation by emigratory behavior possible, namely that of low migrant survival, which is due to the inability of migrants to enter an already occupied local habitat. This inability can also be interpreted as population regulation by immigration. On the other hand, such regulation requires the fraction of emigrating individuals to be rather high; this is not the case here, since the fraction equals the probability of extinction. If, for example, $E = 0.1$, so that 10 percent of Rk young individuals leaves the local habitat, this only diminishes competition among the remaining $0.9Rk$ individuals for k places. Even for a moderate reproductive rate, say $R = 2$, there are still 1.8 young individuals competing for each parental position. Presaturation emigration with such a low fraction of migrants cannot appreciably decrease population density. Some other mechanisms are required to explain presaturation emigration and regulation of population density by such emigration. These are presented in the next section.

8.4. EMIGRATION FROM LOCAL POPULATIONS WITH UNEQUAL RESOURCE PARTITIONING

The effect of unequal resource partitioning on emigration will be presented here applying function $y(x)$, defined in Chapter 2. As in the model presented in section 8.3, nonoverlapping generations are assumed here, as well as an inability of immigrants to enter already occupied local habitats and a number of immigrants large enough to fill all the empty habitats during one generation. Contest competition is also assumed, which, according to the definition given in section 2.5, implies that an individual's resource share $y(x, V)$ is independent of population size N.

Consider a local population of N individuals at the state of equilibrium, without emigration. If in this local population only k $(k < N)$ individuals obtain at least the minimum amount m

of resources, then only these k individuals reproduce, giving a total of $\mathcal{N}$ offspring (Figure 8.2). This is described by equations (2.2) and (2.3), which in a slightly modified form gives

$$\mathcal{N} = h \sum^{x=k} [y(x) - m], \qquad (8.8)$$

where h denotes the conversion coefficient of resources into progeny. According to the definition given in section 2.1, the paramenter k can be interpreted in two ways: (1) as the number of individuals for which $y(x, V) \geq m$, or (2) as the rank of an individual that obtains exactly m resources, so that k can be found by solving equation $y(k, V) = m$.

In section 8.3 it was assumed that the reproductive success of each of the k individuals was the same and was determined

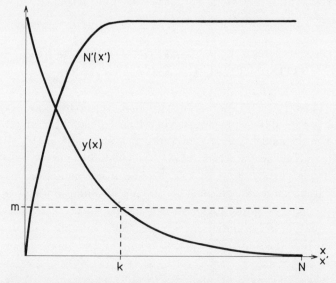

FIGURE 8.2. A hypothetical example of an individual's resource share $y(x)$ as a function of its rank x, for $\mathcal{N}$ individuals, k of which obtain at least m units of resources and are able to produce a total of $\mathcal{N}$ offspring in the next generation; and population size after reproduction, $\mathcal{N}'(x')$ as a function of the threshold rank x', above which emigration occurs, for the same hypothetical example.

by the probability of entering the group of k individuals. With unequal resource partitioning, individuals differ in their reproductive success $f(x)$, which is determined by their rank x according to the equations

$$f(x) = h[y(x) - m], \qquad (8.9)$$

for $x < k$

$$f(x) = 0, \qquad (8.10)$$

for $x \geq k$.

For the sake of simplicity, let us assume that the amount V of resources, and consequently the population size $\mathcal{N}$, is identical in each local population. In order to find a evolutionarily stable strategy of emigration, we have to determine a threshold rank $x = x^*$ with a property such that $f(x^*)$ will be equal to the mean reproductive success of an emigrant F_E. We may expect the individuals with ranks $x > x^*$ to emigrate outside their local populations, and those with ranks $x < x^*$ to refrain from emigration. If $E > 0$, so that some new empty habitats are available and emigrants can reach them, then the mean reproductive success of an emigrant F_E is larger than zero. According to the definition of k, $f(k) = 0$; therefore we may expect $f(x^*) > 0$, and consequently, $0 \leq x^* \leq k$. Thus, not more than k individuals should refrain from emigration. For a more precise determination of x^*, an estimation of the mean reproductive success of an emigrant is required, as presented below.

Below determining the ESS threshold rank (x^*) above which emigration occurs, let us introduce a threshold rank x' $(x' \leq k)$ above which emigration occurs and the state of ecological equilibrium is possible. All individuals whose rank x is larger than the threshold rank x' emigrate from the local population, and they do not enter into any local populations in which there are more than x' individuals. In the next generation, then, the population size $\mathcal{N}'$, determined by the emigration of individuals of ranks $x > x'$, is given by

$$N' = h \sum_{x=x'}^{} f(x). \qquad (8.11)$$

If $N' > x'$, then the population size N' does not change from one generation to the next. If $x' \geq k$, then $N' = N$, as defined by equation (8.8), because contest competition is assumed here, while if $x' < k$, then $N' < N$, because the number k of reproducing individuals is smaller than the maximum possible (Figure 8.2). There are many possible values of x' that are ecologically stable but not necessarily evolutionarily stable.

The population size N' may be also interpreted as the total reproductive success of all x' immigrants that enter a new empty habitat. Since it was assumed earlier that there are always enough immigrants to colonize newly formed local habitats, this implies that the ecological stability of the entire system of local populations is determined by the threshold level x', above which emigration occurs and immigrants do not enter.

For each empty habitat, newly formed with probability E, there are $1/E$ local habitats and $(1/E - 1) = (1 - E)/E$ local populations with $(N' - x')$ individuals emigrating from each. Therefore, for every empty local habitat, there are $(N' - x')(1 - E)/E$ emigrants, and since in this empty habitat they can produce N' offspring, the mean reproductive success F_E of an emigrant is given by

$$F_E(x') = EN'/[(1 - E)(N' - x')]. \qquad (8.12)$$

An evolutionarily stable rank x^* above which emigration occurs is given by equation

$$f(x^*) = F_E(x^*). \qquad (8.13)$$

From this the ESS rank x^*, and so (from equation [8.11]) ESS local population size N^* after reproduction and the proportion of emigrants $(N^* - x^*)/N^*$, can be numerically determined.

This model makes it possible to obtain simple numerical solutions if there is contest competition, so that $y(x)$ is independent of N. An example of such a solution obtained by applying function $y(x, V) = a(1 - a/V)^x$, given by equation (3.11) and described in section 3.6, is shown in Figure 8.3. In relatively stable local habitats, such that $E \leq 0.1$, and with a high reproductive rate, so that there is strong competition for empty places, x^* is close to k; therefore it can be said that only individuals that are unable to survive and reproduce in a given place undertake migration. If there is a high reproductive rate, however, k can be many times smaller than N, which implies that a large proportion $(N - k)/N$ of individuals emigrate outside their local

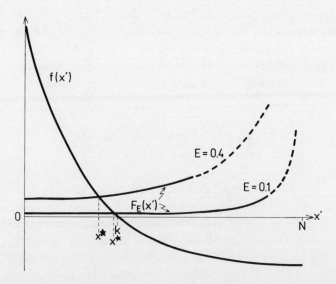

FIGURE 8.3. A numerical example of the determination of the evolutionarily stable threshold rank x^*, above which all individuals emigrate. Individual resource intake $y(x)$ is determined here by the equation $y = V(1 - a/V)^x$, described in section 3.6, and it in turn determines $f(x)$ and $f(x')$. The following set of parameters is applied here: $a = 10$, $V = 300$, $h = 1$, $m = 2$. Function $f(x) = 0$ for ranks $x \geq k$, but in order to show the relative differences between individuals of these ranks, the negative values of $f(x)$ are also shown in the diagram. The above set of parameters yields $k = 47$. The mean reproductive success of emigrants, F_E, is given for two probabilities of extinction of local habitats: $E = 0.1$ and $E = 0.4$, yielding respectively $x^* = 45$ and $x^* = 32$. For higher values of x', the number of immigrants $(1/E - 1)[N(x') - x']$ is smaller than x' and, if so, F_E is no longer determined by equation (8.13). This is shown by the use of broken lines.

habitats. In very unstable local habitats with a high value of E, x^* can be considerably lower than k, which implies that the resources are not completely used up, and the evolutionarily stable local population size N^* after reproduction is smaller than in more permanent local habitats.

In this model of emigration, it is implicitly assumed that at a certain stage of an individual's life, well before reproduction, the future fate of this individual—or, more precisely, its resource intake—is already determined. Furthermore, individuals are able to perceive their future reproductive success from their present status within a local population. This is possible if the differences among individuals are stable, so that a high correlation exists among an individual's present status and its future resource intake. We may also expect that large initial differences among individuals are more strongly correlated with future individual differences than small ones. Therefore intrapopulation variation ought to promote presaturation migration.

8.5. EMIGRATION, AND SCRAMBLE AND CONTEST COMPETITION

The model of emigration presented in the previous section assumes not only unequal resource partitioning, but also contest competition among individuals in local populations. If there were scramble competition with $y(x)$ dependent on N, then emigration of low-ranking individuals would increase the resource intake and reproductive success of those of higher rank, which do not emigrate. This would make the model of the evolution of emigration much more complicated.

Unequal resource partitioning with contest and scramble competition can also be presented in the way proposed by Pulliam and Caraco (1984) for individual fitnesses within groups of different size (Figure 8.4). This is another way of presenting the model described formally in Chapter 2: the amount of resources V (habitat quality) and population (group) size N de-

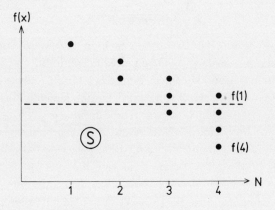

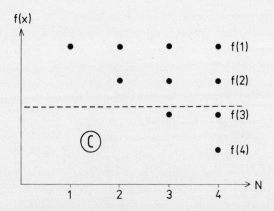

FIGURE 8.4. Examples of reproductive successes $f(x)$ of individuals within a hypothetical local habitat in which either scramble (S) or contest (C) competition occurs. This diagram is based on a similar one by Pulliam and Caraco (1984) for a social group. It presents the reproductive successes of each individual separately as determined by their rank x and their number within a local habitat N. The Allee principle is ignored here, and a resource shortage is assumed, so that with increasing local population size N there is a decrease in the average reproductive success. The average reproductive success of emigrants is given by the horizontal broken line; therefore individuals with a reproductive success below this line should emigrate. If contest (C) competition is assumed, individuals of rank $x = 3$ and $x = 4$ should emigrate, but the reproductive successes of the remaining individuals do not change. If scramble (S) competition is assumed, then in a local habitat with four individuals, three have lower reproductive successes than do emigrants. In spite of this, all three individuals should not emigrate, because after the emigration of the first one, the reproductive successes of the remaining individuals increase, as shown for $N = 3$. These two diagrams show that whereas in contest competition the decision to emigrate or to stay is determined solely by an individual's reproductive success, in scramble competition this decision also depends on the rank and behavior of other individuals in the local population.

termine resource intake $y(x)$, which in turn determines the reproductive success $f(x)$ of an individual of rank x. As assumed in section 8.4 and shown in Figure 8.4(C), in order to make the decision whether to emigrate or to stay, an individual must perceive its present or future reproductive success in its local habitat, and compare it with the average reproductive success of emigrants. For this decision, the presence of individuals of higher rank is of importance, because they determine the reproductive success of lower ranks, but as far as contest competition is concerned, the presence of individuals of lower rank is of no importance. This can be easily seen in Figure 8.4(C): an individual of rank $x = 3$ should emigrate, regardless of whether there are three or four individuals in its local habitat.

An individual undertaking a decision about emigration in a local population with scramble competition is in a quite different situation. Consider a local habitat with four individuals, as shown in Figure 8.4(S). The reproductive success of three individuals in this habitat is lower than the average success of emigrants. If three individuals are present in the local habitat, only one of them has its reproductive success below that of the emigrants. For three of the individuals in the habitat that contains four, the best strategy is to stay and to make the others emigrate. We may expect that the individuals of low rank will not emigrate readily in spite of their low reproductive success in their local habitats. In this way scramble competition may hinder the emigration of a surplus of individuals.

On the other hand, scramble competition does not exclude a possibility of emigration, as long as there are differences in reproductive successes among individuals. Among three individuals with reproductive success below the broken line in a population of four (Figure 8.4[S]), an individual of rank $x = 4$ loses more by staying in the local habitat than does an individual of rank $x = 2$. Figure 8.4 disregards the phenomenon discussed in Chapter 3, that higher density in relation to available resources increases the variation among individuals. If at high densities some individuals have to die before they reproduce (reproductive success equals zero), then the decision that an individual should emigrate is determined by the asymmetric game between two individuals discussed in section 8.1: a weaker opponent should emigrate if its reproductive success within the local habitat is lower than that of emigrants. For individuals of ranks 3 and 4 (Figure 8.4[S]), reproductive success below the broken line is not a sufficient condition for emigration; it is also required that they should have no possibility (or only very low, lower than the emigrant reproductive success) of getting rid of individuals of ranks 1 and 2.

Therefore, it is safe to conclude that although contest competition is not a necessary condition for emigration to occur, it does enhance emigration. As shown in Chapter 6, the mechanism of contest competition, which appears very simple among plants competing for light, is not always clear among animals. In laboratory experiments, like those on Everglades pygmy sunfish discussed in section 6.4, very large differences among individuals alone are not able to bring about contest competition. On the other hand, data collected in the field, like those on gall aphids (section 6.3), suggest the existence of contest competition among these insects. It seems that very often the possibility of emigrating is the condition for contest competition to occur. This conclusion is also supported by data on egg-laying by flour beetles, discussed in section 6.2, in which imagines either were able to emigrate after oviposition or were deprived of this possibility. Simple models of the conflict between two individuals also support this conclusion, since the Dove strategy is more likely to evolve if an individual adopting it has a possibility of gaining something at another time or in another place.

The extent to which contest competition promotes emigration, and to which emigration makes contest competition possible, is therefore an open question. It seems that these two phenomena are acting simultaneously, and at present we have neither the appropriate theoretical models of their evolution nor any good laboratory and field data that would allow us to understand their mechanisms fully. A good appreciation of these two phenomena seems important for understanding the biological basis of population stability.

8.6. FREE AND DESPOTIC DISTRIBUTION
OF ANIMALS

Only individual advantage was considered in this discussion of the evolution of emigration: an individual moves to where its reproductive success is the highest. This is the same principle on which the concept of free and despotic distribution, as defined by Fretwell and Lucas (1970) and further developed by Fretwell (1972), is based. More precisely, when assuming free distribution, an animal is able to move where its fitness is the highest, but when despotic distribution is assumed, the presence of other individuals in the better habitat makes a free choice impossible.

The concept of ideal free distribution can be presented as follows. If population density is low, so that the effects of intraspecific competition are negligible, animals should always choose a better habitat. If density in the better habitat increases, the mean fitness of an individual decreases to the point at which it equals the mean fitness of one in a less crowded but otherwise inferior habitat. An animal will always move to the habitat in which its fitness is higher, and since an increase in the number of animals in the habitat to which animals emigrate lowers their mean fitness, such movements keep densities in both habitats at such a level that the mean fitness is approximately the same in each habitat. Therefore, ideal free distribution predicts that if individuals of a given species occur in two habitats of different quality, their mean fitness in the better habitat is approximately the same as in the worse one.

For despotic distribution, on the other hand, it is assumed that individuals are unable to make the best choice between these two habitats. For example, they may be driven out of the better habitat by high-ranking individuals when the mean fitness in this habitat is still high, and into worse habitats in which the mean fitness is much lower. One can imagine an extreme case of despotic distribution occurring when animals leave a good habitat for the hostile space outside, where their fitness approaches zero. In the case of despotic distribution, the mean fitness in the worse habitat is lower. There are also other reasons why ideal free distribution might be impossible. If, for example, the quality of a habitat declined after individuals had built their nests there, they might stay in this worse habitat, thus diminishing their mean fitness, instead of moving to a less crowded and therefore better place.

How do the concepts of ideal free and ideal despotic distribution relate to the phenomenon of unequal resource partitioning, and to the concepts of scramble and contest competition? Ideal free distribution and ideal despotic distribution are based on the mean fitness of all individuals in a given habitat, not on the distribution of fitnesses. For the case of ideal free distribution, Fretwell (1972, p. 86) assumed that the fitness of each individual should be equal to the mean fitness of all individuals in its habitat. In terms of the model presented in Chapter 2, this implies that equal resource partitioning among population members should be a condition for ideal free distribution. Is this really the case?

An inequality among population members can be due either to the relation between an individual and its environment or to the properties of the individual itself. For example, in the former case, an individual may exhibit higher fitness because of the high quality of its home range or because of poor competitive abilities of its neighbors, while in the latter case its fitness can be due to its size and thus would not change when the individual moves from one habitat to another. If the fitness of an individual is not related to the area in which it lives, we can imagine that animals differ in size and are able to secure different amounts of food; nevertheless they are able to exhibit the ideal free distribution. This was experimentally confirmed by Harper (1982), who studied the distribution of mallards between two different feeding sites on a pond. In his experiments, fitness was measured by the number of food particles taken by a duck, and the quality of a feeding site was determined by the number of food particles supplied to this site per unit of time. The ducks in these experiments were ideally freely distributed with the number of individuals in a site proportional to the number of food particles supplied to this site, and with identical mean numbers of food particles taken by the ducks in both sites. On the other hand, there were large differences in the number of food particles taken by individual ducks, so that some of them took ten times more than others.

The experiments on mallards described above do not entail a real choice of a habitat for an entire season, but rather a short episode of selecting a feeding site. Therefore the differences among individuals were not due to their relations to particular habitats, but to their ability to catch food particles. In the winter, when changes in dominance hierarchy occurred among these ducks and a group of four dominants was formed, their distribution among feeding sites was no longer ideally free. These experiments show that the ideal free distribution does not require equal resource partitioning, and that social hierarchy may prevent the occurrence of this distribution.

The relation between social hierarchy and the distribution between habitats of different quality was analyzed more closely by Pulliam and Caraco (1984), who presented the fitness of each individual within a group graphically, as a function of the quality of local habitat and size of a group inhabiting it. The graphic model applied by these authors can be used to represent both scramble and contest competition, as shown in Figure 8.4; but Pulliam and Caraco used it to represent competition close to contest, so that only the fitnesses of the individuals of low ranks are considerably reduced by increasing group size. When the group size reaches the point at which the fitness of the individual of lowest rank is lower than its possible fitness in another habitat or another group, we expect this individual to leave the group. The average fitness of the individuals of higher ranks, which stay in the group in the better habitat, is much higher than the fitness of those going to a worse habitat or to form a new group. Therefore, it can be said that contest competition leads to despotic distribution.

Contest competition implies that individuals of low ranks are unable to influence the resource intake y of those of high rank; therefore, the fitness of the high-ranking individuals is higher than the mean fitness in the entire habitat. The individuals of low rank emigrate because of their own low fitness, not because of the low mean fitness of the entire population in a given habitat. Consequently, the worse habitat into which individuals of low rank immigrate may cause their fitness to be much lower than the mean fitness in the better habitat they left and in which individuals of high rank stay. Contest competition should therefore bring about despotic distribution. On the other hand, one cannot claim that scramble competition always leads to free distribution. The definition of scramble given in section 6.1 does not exclude a possibility of the emigration of weaker individuals into a suboptimal habitat, in which their fitness is lower.

Other qualifications on the relation between contest competition and despotic distribution should be made. The gall aphids discussed in section 6.3 were used as an example of contest competition, but as Whitham (1980) showed, the mean reproductive success of aphids inhabiting leaves of different sizes was the same, which implies that stem mothers are distributed among leaves in the ideal free way. On the other hand, as shown in section 6.3, competition among aphids within leaves has to be classified as contest competition. If we keep in mind that there are two kinds of habitat within a leaf—the better ones close to the leaf base and the worse ones in more distant parts— then obviously the distribution of aphids among these two habitats on the same leaf is despotic. The character of distribution may change in time. It seems that stem mothers of gall aphids exhibit scramble competition when they select a leaf, and contest competition when they select a place within a leaf. In the same way, hard winter conditions change social hierarchy among mallards and introduce despotic distribution among feeding sites.

Field and Laboratory
Populations of Animals

There are two common study methods in population ecology: (1) counting, collecting, trapping, or marking individuals at a chosen place in the field; or (2) studying a group of animals that are able to reproduce in the laboratory, in confined containers of various kinds. These methods may be applied to simple estimations of population density changing in space and time, as well as to experimental manipulations both in the field and in the laboratory. It is known that these methods may give us a biased picture of natural populations. In this chapter I shall discuss the sources of such bias, which are rarely taken up in ecological textbooks, and which occur both in the field (section 9.1) and in laboratory studies (section 9.2). The effects of confinement on laboratory populations are discussed in sections 9.3 and 9.4, on field populations in section 9.5.

9.1. LIMITATIONS OF FIELD STUDIES

When studies of populations are carried out in the field, time is often limited. Counting individuals, collecting and killing them, or trapping and marking them, are all time-consuming activities. Most of the time is spent not on handling a specimen but on searching for the next one, inspecting traps and moving from one trapping station to another. To collect a sufficient amount of data to allow for statistical inference we need many individuals, and this can only be achieved in areas of extremely high population density. For example, it takes about two hours to search an area of 100 square meters for land snails *Helix pomatia*. In such an area, in places of their maximum density, we may find 25 snails, thus obtaining a considerable amount of data in a day's work. Yet such high densities are rare, and in most other places, densities of this species are ten or a hundred times lower. Population studies in areas of low density take a long time and give us very little information.

Because of the time and labor shortage, field population studies are generally limited to populations at very high densities. Population studies at low densities are extremely rare, and they are carried out under special circumstances—when all the species of a taxonomic group are collected, for example, or when a fluctuating population is monitored in time. But even if such studies are performed, few data are collected, and therefore our knowledge of populations at low densities is almost nonexistent.

The low density of a given species may be due to a low density of its food particles, and therefore we may expect that the time of searching for the next food particle is the most important limitation facing the individuals of this species. Here we may encounter a kind of intraspecific competition that is quite different from that in densely populated areas, where there is no time limitation. The effects of such competition should be similar to those of contest competition: a weaker individual unable to obtain enough food may have a weaker influence on the food intake of stronger ones than in places in which animals are not time-limited. The emigration of such an individual may not increase the amount of food taken by the remaining population members.

When studying a patchily distributed species, we are usually limited to the study of patches that form local habitats, and know very little about migrant survival outside local habitats. If we find a single specimen of a species, the biology of which is less well known, in a certain spot, we usually assume that this spot lies within its natural habitat; the fact that this species is rare is usually explained by the action of some unknown factors limiting its density. Without a detailed knowledge of the ecological requirements and life history of a species, it is usually very difficult to decide whether an individual of this species is in its local habitat or is migrating in hostile space outside its local habitat. Some of the differences between local habitats and hostile space can be even more difficult to detect. For example, host resistance to a parasite makes a host, or the place in which this host is staying, a hostile space, instead of a local habitat. Some extreme and spectacular phenomena, like the presence of many insect species on glaciers or on sea beaches, show that emigrations of animals into hostile space, where the probability of their survival is very low, are common in nature. In other habitats that are not so obviously hostile, it is very difficult to study migrants, because the migrants are rare and stay for a short time.

9.2. ANIMAL POPULATIONS IN THE LABORATORY

The most common way to obtain information about population processes is to conduct a laboratory study of a population in a limited space on a limited amount of food or other resources. A laboratory population is often treated as a sample of a natural one, with the most important elements of its habitat simplified to a degree that allows the experiments to be properly and conveniently carried out.

Two basic conditions should be fulfilled by a habitat appropriate for the experimental population. First, if density is low and the animals are not very old, most of them should be able to survive. This means that the experimental habitat should provide a physical environment and resources of a quality that would enable animals to survive at least to reproduction. Second, the quality of this habitat ought to allow animals to reproduce, at least if population density is low. These two levels of habitat quality, the first making survival possible, the second permitting reproduction, are necessary conditions for the study of population dynamics in the laboratory.

One can imagine a third level of quality of the laboratory habitat, that would not only make survival and reproduction possible, but at which animals would stay without attempting to leave the habitat, if density were low. This third level of habitat quality, based on the behavioral reactions of animals, is usually ignored. Ecologists do not worry whether animals really want to live in a cage or a vial that constitutes their laboratory habitat, since it is almost always confined, with no possibility of emigrating. Laboratory confines are in most cases different from the boundaries of a natural population in the field: these confines prevent emigration, but they do not kill an individual that attempts to emigrate. In the field there is no physical barrier like a glass wall or wire fence; instead, there is another habitat that, when entered, increases the probability of the death of a migrant. Small water pools from which aquatic animals cannot get out are natural habitats in which the confinement is of the same kind as in the laboratory. But even in this special case, if a water pool is larger and spatially heterogeneous, so that it includes both local habitats and the more or less hostile space around them, this is actually different from laboratory habitats. Other environmental islands, such as land surrounded by water, leave open the possibility of migration outside such islands but with a great risk to the emigrating individuals.

When they ignore the third level of habitat quality, ecologists implicitly assume that migratory behavior is of relatively minor importance and that the basic properties of animals, the properties that determine population dynamics, are reproduction and death. These two properties are determined in turn by the available space and resources. In laboratory investigations we are limited not only by time but also by space. For this reason, the concentration of resources in artificial laboratory habitats is many times higher than in the field, and therefore population densities are also much higher. If food is not a limiting resource, overcrowding becomes an important factor modifying animal behavior and the dynamics of laboratory populations, as discussed in sections 7.2 and 7.3.

There are no good reasons why laboratory populations should be kept in confined containers only. Most of the animal species used by laboratory ecologists, such as flour beetles or mice, are pests of food productions. For these species, laboratory conditions do not differ greatly from field conditions, with the exception of confinement. With a little more space and some simple additional arrangements, it is usually possible to attain the third level of quality for laboratory habitats, where at least part of the population members would stay in these habitats, even if they had the possibility of emigrating. Laboratory population studies carried out in such habitats make it possible to approximate the natural phenomena more closely. Short descriptions of the results of such studies in populations of *Hydra* and *Tribolium* are given in sections 9.3 and 9.4, respectively.

9.3. FREE AND CONFINED LABORATORY POPULATIONS OF *HYDRA*

Freshwater polyps of the genus *Hydra* are among the best objects for population studies in the laboratory (Slobodkin 1964). If polyps are placed in small glass vials with frequently changed pond water and are supplied daily with a limited number of small live crustaceans, they reproduce by budding, up to the moment when food begins to run short. Then some individuals stop budding, and later they decrease their body size. The death of an individual in such a population follows a decrease in its body size to the point at which the individual becomes unable to catch a food particle. Population growth is stopped by the shortage of food, and the population size fluctuates with the mean size linearly related to the amount of food supplied (Slobodkin 1964). The mechanism of population regulation in such confined populations is simple: when there is not enough food, more polyps stop budding and die, decreasing the population size in relation to the amount of food supplied, and then allowing more of the remaining polyps to bud and fewer to die.

When one inspects a laboratory population of *Hydra* at high density and limited food, one can easily see that polyps are small, they do not usually bud, and many of them are not fixed to the glass by their pedal disc, but rather float, hanging on the surface film of the water. This floating behavior results from the formation of a bubble of gas beneath the polyp's pedal disc. What is the fate of a floating polyp in a pond or in a lake where only some parts of the bottom may be regarded as an appropriate habitat for *Hydra*? A floating individual leaves its local habitat and can easily be carried away by some horizontal water current to an unknown place, with virtually no possibility of returning. There are no empirical data on the survival of polyps floating on the water surface, but since these polyps are exposed to two kinds of predators—those operating above and below the water surface—one can expect their mortality during the floating episode to be rather high.

Is *Hydra* floating behavior a reaction to food shortage and high population density? Is this reaction variable enough that only a fraction of the polyps will float at a given time? The positive answers to both these questions suggest that facultative emigration via floating may be a mechanism of population regulation in nature. Experimental data reported by Łomnicki and Slobodkin (1966), as well as more extensive studies by Ritte (1969), showed that polyps differ in their reaction threshold, so it is very rarely that either all or none of them float. The density of polyps and the amount of food supplied determine the proportion of floating individuals: well-fed *Hydra* kept at low density in unconditioned water do not float at all, while hungry polyps kept in conditioned water or at high densities float very frequently. Thus *Hydra* floating is a reaction to food shortage and high population density.

179

If floating can act as a mechanism of population regulation, then by removing the floating individuals it should be possible to maintain a laboratory population in a state of equilibrium. This equilibrium should be below the density level determined by food shortage in confined populations. As shown by Łomnicki and Slobodkin (1966) and by Ritte (1969), this is, in fact, the case. An emigration process that was simulated by the removal of floating polyps regulated population density at a level much below that of confined populations. Ritte found that the densities of free populations are 7 percent to 41 percent of the densities of confined populations kept under the same conditions. The oscillations of population density in free populations are much smaller than in confined ones. Polyps in free populations from which floaters are removed are in a much better physiological state: they are heavier, and they bud more frequently. The mean weight of floating polyps that were removed from free populations was close to the mean weight of all polyps from confined populations, but much lower than the mean weight of non-floaters from free populations. The removal of the smaller floating polyps allows the nonfloaters to obtain more food and to bud more frequently.

Since free populations had lower densities and were supplied with food for a short period, the polyps did not have enough time to take all the food particles supplied, so that some were unused. In more dense confined populations, all food particles were immediately taken by the polyps. The reason for this is a low survival of crustaceans of the genus *Artemia* in fresh water, so that feeding time in the laboratory could only last for an hour per day. In confined populations that were more dense, all the *Artemia* were eaten in the first few minutes of feeding.

In confined populations of *Hydra*, the population size was almost linearly related to the amount of food. In free populations, size does not increase linearly with the amount of food, because floating is more frequent at high densities per water volume, due to water conditioning. It is an open question whether such high densities and consequently such high levels of water conditioning may occur in the field. Floating as a reaction to density per water volume may either be an adaptation or a physiological necessity, a by-product of other unknown processes that make polyps float if excessive water conditioning occurs. If this reaction is an adaptation, it can be interpreted as a mechanism that allows polyps to adjust to their future fate within their local populations. A high conditioning level implies that there are many polyps in the immediate surroundings and that they obtain enough food to be able to excrete metabolites intensively into the water. Consequently, the amount of food for individuals in worse positions will further decrease in the future, and therefore emigration from this place by floating is the best strategy.

The possibility of *Hydra* population regulation by floating under laboratory conditions suggests that such regulation may also occur in the field. The experiments described above allow for the following conclusions. First, population regulation by floating is due to unequal resource partitioning among *Hydra* polyps; some individuals are smaller and receive less food, and others are larger, well-fed, and consequently are less likely to float. Second, the equilibrium densities of free populations are much lower than those of confined ones. Applying the symbols introduced in section 8.4, the population size x^* determined by emigration is lower than the size k that can be supported by the available resources. Third, if the population density is regulated by floating, then the food supplied is not completely used. This phenomenon may have a counterpart in nature, where a low density of polyps enables small crustaceans to avoid falling prey to them. Fourth, from what was written in section 8.2, natural local habitats of *Hydra* can be classified as habitats of indeterminate time of persistence, because the emigrating individuals are of the lowest ranks; they are small and hungry.

The partial exploitation of food due to the regulation of population density by emigration can be predicted, assuming contest competition and varied individual resource intake described by equation (3.11), as shown elsewhere (Łomnicki 1978). On the other hand, the competition among *Hydra* polyps is more of the scramble than of the contest type. The simple models of scramble competition given in section 2.2 do not allow for unused resources to be left, because they do not address the limitation of feeding time. Nevertheless, both from simple models of the predator-prey systems and from the experiments on *Hydra*, we may conclude that low population density diminishes the amount of food taken by the entire population. This explains why after the emigration of some of the individuals from a local population, the food left by them cannot be taken by remaining individuals: they simply do not have enough time to do this. This makes emigration more likely, as discussed in section 8.4, because emigration of low-ranking individuals does not increase appreciably the resource intake of those remaining.

Is it reasonable to interpret the results of the laboratory study of free *Hydra* populations as a general rule of population regulation in the field? There are no good reasons why other populations of animals might not be regulated in a similar manner. There is only one serious limitation that should make us careful in drawing general conclusions from the study of *Hydra*: their populations are very often composed of genetically identical individuals, which are members of the same clone. Emigration is much more likely from such clones, as shown in section 8.1. If a population is a clone, then its genetic variance is zero, or close to zero, if the possibility of somatic mutations is accounted for. If so, there is no selection among individuals within a clone to leave the maximum number of progeny within their local habitats, but there is selection among clones to colonize the maximum number of available local habitats. Therefore one may expect the emigration threshold to be at a much lower density than for sexually reproducing individuals, and emigration to occur even if the probability of migrant survival is very low. For this reason, conclusions from the studies of *Hydra* may not be applied generally to sexually reproducing animals.

9.4. FREE AND CONFINED LABORATORY
POPULATIONS OF FLOUR BEETLES

Flour beetles of the genus *Tribolium* reproduce sexually, and therefore the conclusions derived from studies of this group are generally applicable to sexually reproducing species. On the other hand, this generality is diminished by the fact that *Tribolium* habitats are of determinate persistence. Like many other animals used for population studies in the laboratory, *Tribolium* feed on dead plant or animal matter. Whatever the size of the accumulations of their food, which also form their local habitats, these accumulations are limited; therefore, sooner or later all the animals living in these habitats will face the risk of emigration. However, if there is enough food, the flour beetles may stay in their habitats for their whole lifespan, because the adults use the same food as their larvae. Under special circumstances, when new food is continuously added, the local habitats of these beetles may be regarded as habitats of indeterminate persistence.

Flour beetles are good objects of population studies because one can easily arrange for them a habitat of the third level of quality, as defined in section 9.2, from which only some individuals emigrate at high density. The results presented below, concerning free and confined populations of *Tribolium castaneum*, are a part of my own preliminary studies, completed with the help of Dr. H. Warkowska. Laboratory populations of this species were kept for 18 weeks under standard laboratory conditions for *Tribolium*, on standard medium. Population vials were placed in larger vials, and in free populations larvae and adults could migrate one way to the outer vial on a band of paper 13 millimeters wide. Those found in the outer vials were removed every second day.

The number of individuals in each stage was determined every second week at the time when the medium was changed. The results for two populations, a free one and a confined one, kept on 2 grams of medium each, are given in Figure 9.1. The average number of all individuals (including eggs) in the free population

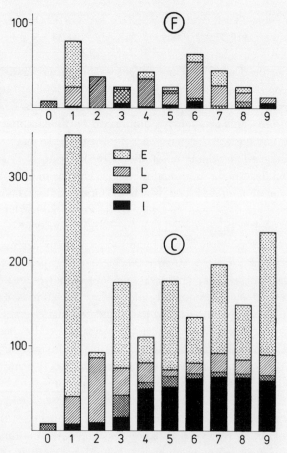

FIGURE 9.1. Number of eggs (E), larvae (L), pupae (P), and imagines (I) in free (F) and confined (C) laboratory populations of *Tribolium castaneum*, counted nine times (every second week) over eighteen weeks.

is about half of that in the confined one. Except for the earliest phase of growth, the free population consists mainly of larvae and pupae; adults and eggs are present in small numbers. In the confined population, after the earliest phase, there are mainly imagines and eggs, but very few larvae and pupae.

Similar results were obtained in free and confined populations kept on 4 and 8 grams of medium.

The biomass of the entire population at the end of the experiment was ten times lower in the free population than in the confined one: 0.0144 g and 0.1572 g, respectively. The amount of medium used, estimated for the last five surveys from the eighth to the eighteenth week of the experiment, was lower in the free population than in the confined one: 11.4 percent and 22.1 percent, respectively (paired comparisons $t = 3.02$, $df = 4$, $P < 0.05$). These experiments show that the mechanism of population regulation in open containers is quite different from that in confined ones. Free populations are regulated by the emigration of larvae—mainly those of the last instar that are preparing for pupation—and by emigration of almost all imagines. During the 18 weeks of the experiment, 66 larvae before pupation and 57 imagines emigrated from the free population. Therefore, the number of eggs laid in free populations is small. On the other hand, eggs are less exposed to being cannibalized by imagines, and they survive much better. In confined populations, the mechanism of regulation is the same as reported for this species by other authors: the cannibalism of eggs, larvae, and pupae, exhibited mainly by imagines (Sokoloff 1974).

Which of these two experimental designs tells us more about mechanisms of population regulation in nature or in free populations of pests, uncontrolled by man? The weaknesses of experiments on free populations are that investigations are limited to local habitats and do not include the hostile space in which imagines search for the next local habitat in which to lay eggs. Therefore they say nothing about the probability of survival and the reproductive success of individuals that have emigrated outside the local experimental habitat. On the other hand, the confinement of the entire population between impenetrable walls is a phenomenon that occurs neither in nature nor in flour and grain storages, which now constitute the environment of this species. Flour and grain are not kept in glass vials, nor in any other containers from which emigration is impossible.

Therefore the phenomena occurring in confined laboratory populations may not have a counterpart in nature.

The experiments on free populations of flour beetles raise new questions. How can the survival and reproductive success of emigrating adults be estimated? It is possible to make laboratory models of metapopulations, with experimentally imposed mortality of migrants, but this mortality would not immediately determine the migration rate. This is because the migration rate has evolved in real metapopulations outside the laboratory. Only after carrying out such experiments for many generations could one expect natural selection to adjust the migration rate to the imposed mortality of migrants. We suppose that the mortality of these beetles outside the laboratory must be very high, because each female is able to lay hundreds of eggs during her life span. At which stage is mortality of these beetles in nature the highest: among eggs and larvae or among imagines? In a free laboratory population the mortality in vials among all stages is rather low, and therefore under natural conditions one can expect either frequent extinctions of such local populations made mainly of larvae or high mortality of migrating imagines, or both. If the mortality of migrating adults is very high, it is surprising that they still exhibit a tendency to leave their local populations after a certain number of eggs per gram of medium has been laid by them or their companions (section 6.2).

9.5. CONFINED POPULATIONS OF ANIMALS IN THE FIELD

If free laboratory populations are able to regulate their density by emigration, and if this is not a laboratory artifact but a real phenomenon, then the differences between free and confined populations should also occur in the field. This can be determined by making an enclosure in the field where the species under consideration lives, and comparing the population within this enclosure with the population outside it. The enclosure made in the field should be large enough to contain all the elements required by animals for survival and reproduction, but small enough to include only a local habitat, not the hostile space outside it. More precisely, the enclosure may include a part of the hostile space but in a much lower proportion than outside the enclosure. If, after a lapse of time, population density and other population parameters inside the enclosure do not differ from those outside, we may conclude that emigration is of no importance for population regulation and that the differences between free and confined populations described in sections 9.3 and 9.4 should be regarded as laboratory artifacts with no relation to the real world.

Confining a field population is usually a difficult and expensive task, although such experiments can be carried out relatively easily for those species that are unable to fly or to climb smooth walls. There are numerous data on field populations studied in enclosures, but unfortunately, these are not accompanied by parallel studies on control populations outside the enclosures, which would make an appropriate comparison posssible. Many authors used enclosures in order to facilitate field investigation, not to study the differences between free and confined populations. For example, Cadwell and Gentry (1965) used an enclosure to study interspecific competition between field mice *Peromyscus polionotus* and house mice *Mus musculus*, while Roughgarden et al. (1984) applied an enclosure to study competition among lizards of the genus *Anolis*. Unfortunately, sufficient data on differences between free and confined populations are not given in their reports.

The first good comparison between free and confined populations in the field was given by Krebs et al. (1969) for two vole species:*Microtus ochrogaster* and *Microtus pennsylvanicus*. These authors studied populations of the two species for two years on grids of 0.8 hectares each, which were either fenced or unfenced. In the fenced grid with no other experimental interferences, the density of *M. pennsylvanicus* attained a peak in late summer, about a year after the beginning of the experiment. This density peak was three times higher than the earlier spring peak of this species in an unfenced grid. In late summer the density of the voles in the fenced grid was four times higher than in the unfenced one. This resulted in considerable overgrazing in the fenced grid. Peak densities of *M. ochrogaster* did not differ between the fenced and unfenced grids, but the pattern of population growth was quite dissimilar: growth was faster in the fenced grid, and the peak density came in winter; in the unfenced grid, the peak density was reached in fall. The introduction of *M. ochrogaster* to an empty fenced grid in spring was followed by fast population growth, with the autumn peak density five times higher than in the control grid. This increase also resulted in overgrazing. Similar experiments were carried out on the vole *Microtus townsendii* (Krebs 1979), and similar results were obtained: high density and overgrazing in the confined population.

The species on which the above experiments were carried out are special ones, since they exhibit cycles of population density. Nevertheless, the phenomenon of high density and overgrazing within fences seems to be a general one. This phenomenon should not occur if the hostile space in which heavy mortality occurs is a part of the fenced area. This prediction is supported by the experiments of meadow voles *M. pennsylvanicus* (Tamarin et al. 1984), in which grids were placed on the edge of woodland, and the fully fenced grid included part of it. Voles trapped within the woodland part of the fenced grid were removed; therefore this part formed a dispersal sink. Four years' study of three grids—fully fenced, unfenced from the side of the woodland, and unfenced—showed no differences in population dynamics due to fencing.

CHAPTER TEN

Spatial and Temporal Heterogeneity and Stability of Ecological Systems

The idea that the natural regulation of animal numbers may be due to dispersal from favorable habitats into unfavorable areas is not a new one. Such a mechanism was proposed by Grinnell (1904), Lidicker (1962, 1975), and Wynne-Edwards (1962) among others, and more recently by Taylor and Taylor (1977, 1978, 1983). There is an important reason, I think, why this idea has not been generally accepted: it is difficult to imagine how natural selection could allow for a large proportion of individuals to disperse, if their survival during the dispersal episode is low. Without these two conditions—a large proportion of dispersing individuals and their low survival—fulfilled in the same populations, emigration cannot act as a mechanism of population regulation. Both our intuition and some theoretical models (sections 8.1 and 8.3) show that either emigration outside local habitats is relatively safe, so that a large proportion of individuals emigrates, or it is dangerous. In the former case emigration cannot regulate population density, and in the latter it could not have evolved by natural selection among individuals. There have been many attempts to solve this dilemma. A solution would be possible if group selection were assumed, as proposed by Wynne-Edwards (1962), but other authors have been trying to tackle this problem within the frame of the Darwinian paradigm of individual selection. In section 8.4, I have tried to show that the solution to this apparent dilemma may

189

lie in accepting the fact of unequal resource partitioning and unequal ranking of individuals within local populations.

If we accept the concept of population regulation by emigration, we may ask why the number of progeny produced is so high that for many individuals the best choice is to emigrate into hostile areas, outside their local habitats. A simple attempt to answer this question is given in section 10.1. In sections 10.2 and 10.3 the concept of spatial heterogeneity is discussed, and the possible consequences for ecosystem stability of population regulation by emigration are presented in section 10.4.

10.1. REPRODUCTION IN SPATIALLY AND TEMPORALLY HETEROGENEOUS ENVIRONMENTS

In section 8.4 I tried to show how unequal resource partitioning among individuals can promote the emigration of a large proportion of population members into hostile areas outside their local habitat, in search of another habitat. Unequal resource partitioning allows for the evolution and maintenance of such emigratory behavior, even if migrant survival is very low. However, it does not explain the overproduction of offspring, which are unable to survive and reproduce in their local habitat and which must face the risk of dispersion.

The question of offspring overproduction is a general one, since it applies not only to animals that are able to choose whether to disperse or to stay, but to all organisms that produce more offspring than are able to survive to maturity. This is one of the fundamental questions of evolutionary ecology (for a review, see Horn and Rubenstein 1984), usually discussed in terms of r and K selection. After the criticism I have raised against the logistic equation in section 1.3, it seems reasonable to propose a different approach to the optimization of reproductive rates.

The simplified model presented below is based on the model of limited population growth in separate discrete places (section

1.4). There are two important simplifications of this model: (1) nonoverlapping generations with generation time equal to a time unit, and (2) low migrant survival, such that the probability of colonization C of an empty place can be defined by following equation:

$$C = (R-1)sp(t),$$

where $p(t)$ denotes the proportion of places occupied at time t, R is the number of offspring produced by an individual occupying a place, and s is migrant survival.

Note that this model assumes implicitly the existence of inequality among population members. An individual that occupies a certain place cannot be displaced by an immigrant arriving in this place, so that the population is divided into two groups: (1) individuals that occupy a place in which they are able to reproduce, and (2) newborn or migrant individuals that are trying either to inherit the place in which their parents live or to find another empty place. Following equations (1.8) and (1.9), the proportion $p(t+1)$ of places occupied by individuals at time $t+1$ is given by the equation

$$p(t+1) = p(t) + [1-p(t)](R-1)sp(t-p(t)E, \qquad (10.1)$$

where E denotes the probability of extinction of an individual's place, which is equivalent to its death. From equation (10.1), the nontrivial equilibrium point at which $p(t+1) = p(t) = p_e$ is given by

$$p_e = 1 - E/[s(R-1)]. \qquad (10.2)$$

The fitness F of an individual in this model can be measured in its ability to increase the proportion of occupied places, and therefore it can be given by the formula

$$F = p(t+1)/p(t), \qquad (10.3)$$

which after substituting (10.1) yields

$$F = 1 - E + s(R - 1)[1 - p(t)]. \tag{10.4}$$

By definition, at the state of equilibrium, equation (10.4) is reduced to $F = 1$.

To determine the ESS number of offspring, R, imagine a mutant that produces a different number of offspring, $(R + b)$, instead of R. An increase b in the number of offspring may have adverse effects on the mutant's success. Let us assume that the survival of the established individuals and the colonization ability of migrants decrease linearly with an increase in the number of offspring. The fitness F_M of such a mutant if given by

$$F_M = (1 - E)(1 - cb) + s(R + b - 1)[1 - p(t)](1 - db), \tag{10.5}$$

where c and d are coefficients relating a decrease in the survival of established individuals and in the colonization ability of migrants, respectively, to an increase b in the number of offspring. Let us also assume that these decreases are small, such that $|cb| < 1$ and $|db| < 1$. By setting $p(t) = p_e$, as defined by (10.2), and $F = F_M$, as defined by (10.4) and (10.5), the evolutionarily stable number R^* of offspring is given by

$$R^* = 1 + E(1 - db)/[c(1 - E) + dE] \tag{10.6}$$

With a small change in the number of progeny (i.e., b approaching zero), and with the assumption that this change has the same effect on migrating and settled individuals so that $c = d$, equation (10.6) reduces to

$$R^* = 1 + E/c,$$

which implies that the ESS number of offspring is linearly related to prereproductive mortality E. Imagine an absolutely stable environment, with no mortality before the reproduction episode $(E=0)$. The best strategy of any plant or animal in such an environment is to produce one very strong offspring, which would be able to replace its parent. With increasing mortality, so that new places are becoming available, the ESS reproductive rate R^* increases linearly with the proportion of these new places, as expressed by parameter E. This increase is arrested by the adverse effects of higher reproduction b, expressed by c and d. An adverse effect d of higher reproduction on colonization ability affects R, but migrant survival s itself does not affect it. If for some reason these adverse effects are not very strong, we may expect a very high reproductive output in relation to the number of available places, and consequently, high juvenile mortality.

If the natural environments of plants and animals were invariable in time, then natural selection would favor the production of a few very strong offspring with strong competitive ability. Spatial and temporal heterogeneity favors those individuals which are able to find newly formed habitats away from the place in which they were born. Searching for a new favorable place can be done in two different ways: either by active movement over a large area to seek out empty and favorable spots, or by a high reproductive rate and random dispersion. It seems that this first method is not much more efficient than the second: high reproduction and dispersion alone require no special adaptation for active movement, and they make it possible to search for empty places over large areas. To cope with environmental heterogeneity by the overproduction of offspring seems to be the most common method among all organisms. This is true not only for those species in which hundreds and thousands of offspring are produced, but also for large mammals that exhibit relatively low reproductive rates.

The human mind is accustomed to various artificial systems that have been designed and constructed by man. Such systems are made to perform certain functions with the maximum efficiency and minimum use of energy or raw materials. When studying ecological systems, we are inclined to search for analogies with the more familiar man-made systems, but there are few such analogies. Neither populations nor ecosystems were designed by humans, and although a single individual can be regarded as an entity with goals to aim for—reproduction and survival—no such goals exist at the level of the population or of higher ecological entities. An individual can be efficient, but there is no reason for populations or ecosystems to be efficient. The soil is full of seeds that will never germinate, the air is full of all kinds of propagulae that will never find a place to reproduce. The overproduction of offspring can be explained as an optimal strategy for an individual; for a population or an ecosystem this is an enormous waste. This overproduction allows for the existence of large reserves, so that most perturbations that lower population size or destroy a population completely can be easily counterbalanced by immigration or by available reserves. This is a kind of stability quite different from that encountered in man-made systems.

10.2. SPATIAL MICROHETEROGENEITY

In section 8.2, spatial heterogeneity was regarded as a factor that divides a certain space into local habitats and hostile areas outside. However, such spatial division does not exhaust all the possible kinds of spatial heterogeneity. In nature we can also expect large spatial differences within a local habitat. A single leaf seems much more heterogeneous than a vial containing the medium that supports a laboratory population of fruit flies, and a fluid medium for a protozoan culture, which is constantly mixed, is even more homogeneous. Let us call the spatial heterogeneity expressed in differences between local habitats and the remaining space "macroheterogeneity," and let us call the heterogeneity within local habitats "microheterogeneity."

When they discuss spatial heterogeneity, contemporary biologists are primarily interested in how the heterogeneity is related to genetic variability. For example, Bell (1982) sees spatial heterogeneity as the most important factor among those which make the evolution and maintenance of sexuality possible. I will ignore here the relation between spatial heterogeneity and genetic variability, but it is important to note that if a space is heterogeneous and the genetic make-up of individuals differs, we cannot expect all individuals to find those places best fitted to their genetic make-up. The more or less random dispersion of animals and plant seeds should lead to a great variability in their reproductive success, depending on the place in which they happen to settle and on their genetic background. This is valid not only for the dispersion outside local habitats, but also for that within local habitats. Spatial microheterogeneity should bring about differences in resource partitioning and reproductive success among individuals.

The relation between individual success and spatial microheterogeneity seems so obvious that few investigations have been made to confirm it experimentally. This relation is, as a matter of fact, obvious for plants, and when one inspects patterns of germination and growth of plants, one can easily see how local conditions influence their reproductive success. Harper (1977) listed many experimental data concerning the influence of spatial heterogeneity on plant populations.

For animals, the relation between individual success and spatial heterogeneity depends on whether there is contest or scramble competition. If each individual is free to use the best places within its local habitat, spatial microheterogeneity may not cause an increase of individual phenotypic variation. This may be seen especially easily under laboratory conditions in which animals are kept in small containers and given food of high quality. If an individual can reach each grain of a better food very easily, and if these grains are not defended by other individuals, then the existence of such grained food distribution does not increase the phenotypic variation among individuals.

Preliminary experiments carried out by Ciesielska (1985) on *Tribolium confusum* failed to confirm the relation between spatial microheterogeneity and individual variability. The larvae of this species were reared either on a homogeneous medium made of fine-grained flour and yeast or on flour mixed with unground pieces of yeast, as well as on a medium made of yeast placed in a layer on the top of the flour. It is known that yeast enriches the medium; therefore the medium in the immediate vicinity of a grain of yeast should be of higher quality. Despite this, such experimental heterogeneity has not significantly changed the coefficients of variation of pupal weights. It seems that larvae can easily reach every place in the vial and compete with their companions in each place on equal terms.

These experiments do not exclude the possibility that under different conditions, environmental microheterogeneity may bring about individual variation. Theoretically, even if the grains of better resources are not defended, an individual that has happened to get a better grain of food by accident may have a greater chance of getting another better piece of food. This was discussed in section 3.3. Therefore, even under scramble competition, we may expect microheterogeneity to promote individual variability. This may be especially pronounced if food is more scarce under natural conditions than it is in the laboratory. Under such circumstances an individual may not be able to reach all the best food particles, all the best shelters, or other best elements of a heterogeneous environment that would increase its reproductive success. Unfortunately I know of no good experimental data that could confirm this prediction.

If individuals are more or less stationary, and if resources are defended, as among gall-aphids (section 6.3), then spatial heterogeneity undoubtedly brings about individual variation. Under such circumstances we can expect a strong correlation between spatial microheterogeneity and the variability in individual success.

If we accept the existence of a relation between environmental microheterogeneity and individual variability, then consequently we should accept the relation between microheterogeneity and population stability. Furthermore, as shown in section 8.4, individual differences make emigration from local populations more likely; thus microheterogeneity should promote emigration outside local habitats into hostile areas. On the other hand, macroheterogeneity, which makes the areas outside local habitats hostile, results in a low survival of migrants and therefore allows for population regulation by emigration. Both micro- and macroheterogeneity, then, have a stabilizing effect on natural populations.

10.3. DIRECT RELATIONS BETWEEN SPATIAL HETEROGENEITY, INDIVIDUAL VARIABILITY, AND STABILITY

The relation between spatial heterogeneity and population stability can be studied directly without the explicit introduction of unequal resource partitioning. Hassell and May (1973), investigating the conditions for stability in theoretical models of insect host-parasite systems, have found that the aggregation of parasites, as well as spatial and temporal asynchrony, contribute to stability.

De Jong (1979) has proposed a model of a single-species population dispersed over patches of resources, so that competition among individuals occurs within patches. She has shown that such dispersion enhances population stability. Assuming the same amount of resources in every patch but a variable number of individuals, one obtains various densities of these individuals in relation to the available resources. The competition in some patches is high with low survival, while in other patches competition is low with a large proportion of survivors. Therefore, an increase in the size of the entire population would not produce severe mortality throughout the entire area, which might eventually lead to population fluctuation; severe mortality would occur only in some of the patches. The entire population is not made of identical individuals that react identically to the increasing population density, but of groups that differ in this reaction.

Hassell and May (1985) have reviewed the progress made recently along the line of reasoning presented above. Starting from the model by de Jong (1979) of a single-species population, they presented similar models concerning competing species and host-parasitoid interaction, and they reviewed the influence of patchiness on other systems, like predator-prey, parasite-host, and disease-host. If the negative binomial distribution, with variable clumping of individuals among patches, is applied to de Jong's model, it can be shown that clumping enhances the stability of the entire population. In all models reviewed by Hassell and May (1985), the clumped distribution, and other mechanisms that bring about differences among individuals, enhance stability. A stabilizing effect can be due, for example, to the differences in host susceptibility. A varying resistance to parasites brings about differences in the probabilities of survival both for the host and among the parasites. Therefore differences in susceptibility bring about similar effects, such as differences in resource partitioning: some individuals survive and others die, without large fluctuations in density.

In Chapter 2, attempts are made to deduce the function of survival $k(\mathcal{N})$ by applying the function of individual resource intakes $y(x, V, \mathcal{N})$, as determined by the initial number of individuals $\mathcal{N}$ and the amount of resources V for the entire population. De Jong (1979), as well as Hassell and May (1985), do not deduce the function of survival from unequal resource partitioning, but rather they assume it, as a basic property of the system. They use, for example, the negative exponential according to the formula $k/\mathcal{N} = \exp(-d\mathcal{N})$, where d is a constant coefficient. Therefore the models presented and reviewed by Hassell and May (1985) represent a different approach to the problems of inequality, heterogeneity, and stability. These authors do not explicitly define resource partitioning among individuals and the consequences for stability, as it was done in

Chapter 2. Instead, they describe stability as a result of processes that, after close inspection, appear to bring about unequal resource partitioning. Their approach is different, but the results obtained seem to be quite similar.

10.4. SPATIAL HETEROGENEITY AND
SPECIES DIVERSITY

Ecological space is more or less heterogeneous and is inhabited by different communities of plants and animals. There are various causes of spatial heterogeneity, some of them physical, but the most obvious causes are the plants and animals themselves: their species diversity and their occurrence as discrete units. Irrespective of what may be the cause and what may be the effect here, spatial heterogeneity and species diversity are strongly correlated: more diverse communities occupy more heterogeneous areas. This correlation is obvious for most ecologists, but there are no good and precise data to support it. I think this is mainly because, although species diversity can be easily measured by the number of species or by more sophisticated indices that take account of the abundance of each species, there are no such standard measures for spatial heterogeneity. Species diversity is a clearly defined and easily measured property of natural communities; spatial heterogeneity is a feature that can be easily seen but is very difficult to define precisely and to measure in the field.

From the point of view of a single-species population, spatial heterogeneity makes high reproductive succcess possible in some parts of an area, while the probability of survival and consequently of leaving progeny remains low in other parts of the same area. The concept of spatial heterogeneity, however, seems to include more than just the large differences in the quality of the environment. Imagine a community of 100 species of monophagous insects. If this community occupies an area in which all the food plants are represented, each in fairly low density, so that the most abundant plant species does not exceed 10 percent of the total plant biomass, such an area is heterogeneous; but if the biomass of the most abundant plant species is 95 percent of the total, leaving 5 percent to the remaining species, such an area seems to be much less heterogeneous. Heterogeneity includes, therefore, not only large spatial differences of ecological conditions for a given single species, but also a low proportion of spots favorable for this species within the area. From what has been said above, spatial heterogeneity for the community of monophagous herbivores can be measured by plant species diversity. On the other hand, food may not be the only important element of the environment for these insects. Some microclimatic conditions as well as the action of predators may exclude a herbivore from its food plants. If so, then some indices of the aggregation of the herbivore species might be a better estimate of the spatial heterogeneity for this species than the species diversity of its food plants would be. If spatial aggregation can be used as a measure of environmental heterogeneity, then it would be interesting to find out whether diversity of the entire community is correlated with the spatial aggregation of many of the species inhabiting it.

Defining the measure of spatial heterogeneity is a complicated problem that lies beyond the scope of this book. It seems possible to define it for one species, and there are good reasons to suppose that the spatial heterogeneity of one species should be correlated with the spatial heterogeneity for many other species. The distribution of the individuals of various species in a given area is usually not random; rather, the presence of one species is either positively or negatively correlated with the presence of other species in its immediate vicinity. Therefore, an area heterogeneous for one species should also be heterogeneous for many other species inhabiting it. If so, it seems justifiable to use the concept of the heterogeneity of an entire community or an entire ecosystem, even if at present we do not have any precise methods of measuring it. We have to estimate spatial heterogeneity either for some simple cases or by applying indices that are approximate measures.

10.5. SPATIAL HETEROGENEITY AND ECOSYSTEM STABILITY

Imagine a heterogeneous area with many local habitats for 100 different species. The local habitats of some of these species overlap and some exclude each other, but since there are 100 species, we may suppose that only a small fraction of this space is favorable for any particular one. If, according to the model presented in section 10.1, offspring production of each species in each local habitat is higher than this habitat can support, many will migrate to other local habitats. Unequal resource partitioning makes emigration evolutionarily possible even if migrant survival is very low (section 8.4). If only a small fraction of the area is suitable for a given species, we can expect high-mortality before the migrants can reach other habitats. This can keep 100 different species at relatively low densities.

Now imagine what would happen if for some reason there was a considerable reduction of spatial macroheterogeneity, so that only five different species could find suitable local habitats in this area. The local habitats for these species would be much larger; they might even cover the entire space. If surplus progeny were still produced and migrants moved outside the place in which they were born, they would not enter a hostile environment, but rather the same or another similarly crowded local habitat. Under such circumstances, the homogeneity of the space has the same effect as an enclosure: it makes the regulation of population density by emigration impossible. A reduction of spatial microheterogeneity may also result in smaller differences among members of a local population, which in turn may diminish the rate of emigration from the local population.

201

The relation between species diversity and ecosystem stability is one of the most debated problems of ecology. Attempts are being made to find the mechanisms that relate these two phenomena. A review of these attemps is given by May (1974). Here I do not intend to solve the question of species diversity and ecosystem stability, but rather to suggest a different approach to this question. Among the following three phenomena:

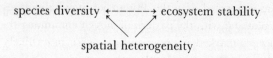

species diversity ⟵－－－－－⟶ ecosystem stability

spatial heterogeneity

most of the theoretical work has previously been devoted to the direct relation between diversity and stability. Spatial heterogeneity has been seen only as a condition for high species diversity or as its by-product, without a direct relation to ecosystem stability. I would like to postulate that there is possibly a direct relation between spatial heterogeneity and ecosystem stability. If we accept that both macro- and microheterogeneity have stabilizing effects on a single population, it should follow that they have stabilizing effects on communities and ecosystems as well.

There are some serious difficulties in investigating stability as related to heterogeneity. As discussed in section 10.4, spatial heterogeneity has not been defined clearly and is difficult to measure. On the other hand, the difficulties we encounter when studying spatial heterogeneity are relatively small when compared with the difficulties encountered when attempting to measure the parameters of a community matrix in the simplest ecosystems. An important advantage of the proposed approach is that we are able to study how the stability of a single population relates to spatial heterogeneity; we do not have to study the entire ecosystem with all its possible complex interrelations.

Some studies within this field seem very promising, although the picture they present is more complicated than that of the simple concept presented above, which relates stability to heterogeneity. Bach (1980), studying the density of striped cucumber beetles *Acalymma vittata* in monocultures and polycultures at two different densities of food plants, has found that the density was 10 to 30 times greater in monocultures. She was not able to attribute this to plant density alone, since the density of the plant host was the same in mono- and polycultures. She has also excluded the adverse effects of other plant species that could occur in polycultures. The only explanation Bach could find was connected with the amount of time individuals spend in one plot and with the pattern of movement among the plots. The beetles marked in monocultures were more likely to be found later on the same plot than those marked in polycultures. It seems that the mechanism of emigration is as follows: after leaving its place, a beetle may come across either its food plant or another plant. If it comes across a food plant, it stays on it, but if it comes across another plant, it moves further, leaving the plot altogether.

The mechanism of the migration of these beetles is more complicated than the mechanisms proposed by simple models from Chapter 8. Migrating individuals can easily enter a habitat that is as crowded as the place they have left. They do not look for an empty habitat, but for an appropriate one, and they stay there even if it is already occupied. Bach's study does not show that the mortality of migrants in polycultures within the same plot is higher than in monocultures, but since dispersal in a polyculture is more often followed by dispersal outside the plot, this may eventually lead to high mortality of migrants far from their local habitats.

The classic models of community ecology are concerned with homogeneous space in which individuals that are identical within a species but that differ between species interact with each other, forming more or less stable systems. After we reject the ideas of the identity of individuals within a species and of the homogeneity of ecological space, a different picture of natural communities emerges, in which individual variation, spatial heterogeneity, behavioral interaction among individuals, and migratory behavior do determine the dynamics and stability of ecological systems. This picture is quite different from that proposed by many theoretical ecologists during the last half-century, but it is not far removed from some concepts developed by field ecologists. The ideas of Grinnell (1904), Errington (1946), and Wynne-Edwards (1962), as well as those developed more recently by behavioral ecologists (Krebs and Davies 1984), seem to fit reasonably well with the picture advocated here.

References

Aitchison, J., and J.A.C. Brown. 1957. *The Lognormal Distribution*. Cambridge University Press, Cambridge.

Allee, W. C., A. E. Emerson, O. Park, T. Park, and K. P. Schmidt. 1950. *Principles of Animal Ecology*. W. B. Saunders, Philadelphia.

Bach, C. E. 1980. Effects of plant density and diversity on the population dynamics of the specialist herbivore, the striped cucumber beetle, *Acalymma vittata* (Fab.). *Ecology* 61: 1515–1530.

Bacon, P. J. 1982. Population dynamics: models based on individual growth, resource allocation and competitive ethology. *Merlewood Research and Development Paper*, no. 88, Institute of Terrestrial Ecology, Grange-over-Sands, Cumbria, England.

Begon, M. 1984. Density and individual fitness; asymmetric competition. In B. Shorrocks, ed., *Evolutionary Ecology*, 175–194. Blackwell, Oxford.

Bell, G. 1982. *The Masterpiece of Nature: The Evolution and Genetics of Sexuality*. University of California Press, Berkeley.

Bellows, T. S., Jr. 1981. The descriptive properties of some models for density dependence. *J. Anim. Ecol.* 50: 139–156.

Brace, R. C., and J. Pavey. 1978. Size-dependent dominance hierarchy in the anemone *Actinia equina*. *Nature* 273: 752–753.

Braun-Blanquet, J. 1932. *Plant Sociology: The Study of Communities*. McGraw-Hill, New York.

Bryant, D. M. 1978. Establishment of weight hierarchies in the broods of house martin *Delinchion urbica*. *Ibis* 120: 16–26.

Cadwell, L. D., and J. B. Gentry. 1965. Interactions of *Peromyscus* and *Mus* in a one-acre field enclosure. *Ecology* 46: 189–192.

Charnov, E. L., and J. P. Finerty. 1980. Vole population cycles: a case for kin-selection. *Oecologia* 45: 1–2.

Ciesielska, M. 1985. "Zroznicowanie fenotypowe poczwarek *Tribolium confusum* w niejednorodnych siedliskach." M. Sc. thesis, Jagiellonian University, Krakow.

Clutton-Brock, T. H., F. E. Guinness, and S. D. Albon. 1982. *Red Deer, Behavior and Ecology of Two Sexes*. Chicago University Press, Chicago.

Cohen, M. N., R. S. Malpass, and H. G. Klein, eds. 1980. *Biosocial Mechanisms of Population Regulation*. Yale University Press, New Haven.

Colinveaux, P. A. 1973. *Introduction to Ecology*. Wiley, New York.

Comins, H. N., W. D. Hamilton, and R. M. May. 1980. Evolutionarily stable dispersal strategies. *J. Theor. Biol.* 82: 205–230.

Cooper, W. S. 1981. Natural decision theory: a general formalism for the analysis of evolved characteristics. *J. Theor. Biol.* 92: 401–415.

Cooper, W. S., and R. H. Kaplan. 1982. Adaptive "coin-flipping": a decision-theoretic examination of natural selection for random individual variation. *J. Theor. Biol.* 94: 135–151.

Dawkins, R. 1982. *The Extended Phenotype: The Gene as the Unit of Selection*. Freeman, Oxford.

Dawson, P. S. 1975. Directional versus stabilizing selection for development time in natural and laboratory populations of flour beetles. *Genetics* 80: 773–783.

De Jong, G. 1976. A model of competition for food. I. Frequency-dependent variabilities. *Amer. Natur.* 110: 1013–1027.

De Jong, G. 1979. The influence of the distribution of juveniles over patches of food on the dynamics of a population. *Netherl. J. of Zool.* 29: 33–51.

Dewsbury, D. A. 1982. Dominance rank, copulatory behavior, and differential reproduction. *Quart. Rev. Biol.* 57: 135–159.

Donald, C. M. 1951. Competition among pasture plants. I. Intra-specific competition among annual pasture plants. *Austral. J. Agricul. Res.* 21: 355–377.

Errington, P. L. 1946. Predation and vertebrate populations. *Quart. Rev. Biol.* 21: 147–177, 221–245.

Falconer, D. S. 1981. *Introduction to Quantitative Genetics.* Longman, London.

Fleischer, R. C., R. F. Johnson, and W. J. Klitz. 1983. Allozymic heterozygosity and morphological variation in house sparrows. *Nature* 304: 628–630.

Fretwell, S. D. 1972. *Populations in a Seasonal Environment.* Princeton University Press, Princeton, N.J.

Fretwell, S. D., and H. L. Lucas, Jr. 1970. On territorial behavior and other factors influencing habitat distribution in birds. I. Theoretical development. *Acta Biotheor.* 19: 16–36.

Fujii, K. 1975. A general simulation model for laboratory insect population. I. From cohort of eggs to adult emergences. *Res. Popul. Ecol.* 17: 85–133.

Gill, D. E. 1979. Density dependence and homing behavior in adult red-spotted newts *Notophthalmus viridescens* (Rafinesque). *Ecology* 60: 800–813.

Gilpin, M. E. 1975. *Group Selection in Predator-Prey Communities.* Princeton University Press, Princeton, N.J.

Gliwicz, J. 1979. Struktura wiekowa a organizacja socjalna populacji. *Wiadomosci Ekologiczne* 25(2): 9–17.

Gould, S. J. 1980. *The Panda's Thumb.* Norton, New York.

Grant, P. R. 1978. Dispersal in relation to carrying capacity. *Nat. Acad. Sci. (USA) Proc.* 75: 2854–2858.

Grinnell, J. 1904. The origin and distribution of the chestnut-backed chickadee. *Auk* 21: 264–282.

Gurney, W.S.C., and R. M. Nisbet. 1979. Ecological stability and social hierarchy. *Theor. Popul. Biol.* 14: 48–80.

Gurney, W.S.C., R. M. Nisbet, and J. H. Lawton. 1983. The systematic formulation of tractable single-species population models incorporating age structure. *J. Anim. Ecol.* 52: 479–495.

Gustafsson, L. 1985. Lifetime reproductive success and heritability: empirical support of Fisher's fundamental theorem. In "Fitness Factors in the Collared Flycatcher *Ficedula albicollis*," Ph.D. thesis, University of Uppsala.

Hairston, N. G., F. E. Smith, and L. B. Slobodkin. 1960. Community structure, population control, and competition. *Amer. Natur.* 94: 421–425.

Hamilton, W. D., and R. M. May. 1977. Dispersal in stable habitats. *Nature* 269: 578–581.

Harper, D.G.C. 1982. Competitive foraging in mallards: "ideal free" ducks. *Anim. Behav.* 30: 575–584.

Harper, J. L. 1977. *Population Biology of Plants*. Academic Press, London.

Hassell, M. P. 1975. Density dependence in a single species population. *J. Anim. Ecol.* 44: 283–295

Hassell, M. P. 1978. *The Dynamics of Arthropod Predator-Prey Systems*. Princeton University Press, Princeton, N.J.

Hassell, M. P., and R. M. May. 1973. Stability in insect host-parasite models. *J. Anim. Ecol.* 42: 693–726.

Hassell, M. P., and R. M. May. 1985. From individual behavior to population dynamics. In R. Sibley and R. Smith, eds., *Behavioural Ecology: Ecological Consequences of Adaptive Behaviour*, 3–32. Blackwell, Oxford.

Hewlett, P. S., and R. L. Plackett. 1979. *An Introduction to the Interpretation of Quantal Responses in Biology*. University Park Press, Baltimore.

Heywood, J. S., and D. A. Levin. 1984. Genotype-environment interactions in determining fitness in dense, artificial populations of *Phlox drummondii*. *Oecologia* 61: 363–371.

Holling, C. S. 1959. Some characteristics of simple types of predation and parasitism. *Can. Entomol.* 91: 385–398.

Horn, H. S., and D. L. Rubenstein. 1984. Behavioral adaptations and life history. In J. R. Krebs and N. B. Davies, eds. *Behavioural Ecology: An Evolutionary Approach*, 279–298. Blackwell, Oxford.

Hubbell, S. P., and P. A. Werner. 1979. On measuring the intrinsic rate of increase of populations with heterogeneous life histories. *Amer. Natur.* 113: 277–293.

Hughes, T. P. 1984. Population dynamics based on individual size rather than age: a general model with a reef coral example. *Amer. Natur.* 123: 778–795.

Ingram, C. 1959. The importance of juvenile cannibalism on the breeding biology of certain birds of prey. *Auk* 76: 218–226.

Jagers, P. 1975. *Branching Processes with Biological Applications.* Wiley, London.

Kaplan, R. H., and W. S. Cooper. 1984. The evolution of developmental plasticity in reproductive characteristics: an application of the "adaptive coin-flipping" principle. *Amer. Natur.* 123: 393–410.

Kimmel, M. 1986. Does competition for food imply skewness? *Math. Biosci.* 80: 239–264.

Kira, T., H. Ogawa, and N. Sakazaki. 1953. Intraspecific competition among higher plants. I. Competition-yield-density interrelationship in regularly dispersed populations. *J. Inst. Polytech., Osaka City Univ.*, ser. D, 4: 1–16.

Knapton, R. W., and J. R. Krebs. 1974. Settlement pattern, territory size, and breeding density in the song sparrow (*Melospiza melodia*). *Can. J. Zool.* 52: 1413–1420.

Kot, M., and W. M. Schaffer. 1984. The effects of seasonality on discrete models of population growth. *Theor. Popul. Biol.* 26: 340–360.

Koyama, H., and T. Kira. 1956. Intraspecific competition among higher plants. VIII. Frequency distribution of individual plant weight as affected by interaction between plants. *J. Biol., Osaka City Univ.* 7: 73–94.

REFERENCES

Kozłowski, J. 1980. Density dependence, the logistic equation, and r- and K-selection: a critique and an alternative approach. *Evol. Theor.* 5: 89–101.

Krebs, C. J. 1979. Dispersal, spacing behaviour, and genetics in relation to population fluctuations in the vole *Microtus townsendii*. In U. Halbach and J. Jacobs, eds., *Population Ecology*, 61–77. Gustav Fischer Verlag, Stuttgart.

Krebs, C. J., B. L. Keller, and R. H. Tamarin. 1969. *Microtus* population biology: demographic changes in fluctuating populations of *M. ochrogaster* and *M. pennsylvanicus* in Southern Indiana. *Ecology* 50: 587–607.

Krebs, J. R., and N. B. Davies, eds. 1984. *Behavioral Ecology: An Evolutionary Approach*. Blackwell, Oxford.

Krebs, J. R., and C. Perrins. 1978. Behaviour and population regulation in the great tit (*Parus major*). In F. J. Ebling and D. M. Stoddart, eds., *Population Control by Social Behaviour*, 23–47. Institute of Biology, London.

Laskowski, R. 1986. Survival and cannibalism in free and confined populations of *Tribolium confusum* (Duval). *Ekol. Pol.* 34: 723–735.

Le Cren, E. D. 1973. Some examples of the mechanisms that control the population dynamics of salmonid fish. In *Mathematical Theory of the Dynamics of Biological Populations*, 125–135. Academic Press, London.

Lerner, I. M. 1954. *Genetic Homeostasis*. Oliver and Boyd, Edinburgh.

Leslie, P. H. 1945. The use of matrices in certain population mathematics. *Biometrika* 33: 183–212.

Levins, R. 1968. *Evolution in Changing Environments*, Princeton University Press, Princeton, N.J.

Levins, R. 1970. Extinction. In M. Gerstengaber, eds., *Some Mathematical Problems in Biology*, Lectures of Mathematics in Life Science, 2: 75–108. American Mathematical Society, Providence.

Lewontin, R. C. 1974. *The Genetic Basis of Evolutionary Change*. Columbia University Press, New York.

REFERENCES

Lidicker, W. Z., Jr. 1962. Emigration as a possible mechanism permitting the regulation of population density below carrying capacity. *Amer. Natur.* 96: 29–33.

Lidicker, W. Z., Jr. 1975. The role of dispersal in the demography of small mammals. In F. B. Golley, K. Petrusewicz, and L. Ryszkowski, eds., *Small Mammals: Their Production and Population Dynamics*, 103–128. Cambridge University Press, Cambridge.

Lloyd, J. A. 1980. Interaction of social structure and reproduction in populations of mice. In M. N. Cohen, R. S. Malpass, and H. G. Klein eds., *Biosocial Mechanisms of Population Regulation*, 3–21. Yale University Press, New Haven, Ct.

Łomnicki, A. 1978. Individual differences between animals and natural regulation of their numbers. *J. Anim. Ecol.* 47: 461–475.

Łomnicki, A. 1980a. Group selection and self-regulation in animal populations. *Ekol. Pol.* 28: 543–555.

Łomnicki, A. 1980b. Zroznicowanie osobnikow a regulacja zageszczenia populacji. *Wiadomosci Ekologiczne* 26: 361–390.

Łomnicki, A. 1980c. Regulation of population density due to individual differences and patchy environment. *Oikos* 35: 185–193.

Łomnicki, A., and J. Krawczyk. 1980. Equal egg densities as a result of emigration in *Tribolium castaneum*. *Ecology* 61: 432–437.

Łomnicki, A., and J. Ombach. 1984. Resource partitioning within a single species population ad population stability: a theoretical model. *Theor. Popul. Biol.* 25: 21–28.

Łomnicki, A., and L. B. Slobodkin. 1966. Floating in *Hydra littoralis*. *Ecology* 47: 881–889.

May, R. M. 1972. On relationships among various types of population models. *Amer. Natur.* 107: 46–57.

May, R. M. 1974. *Stability and Complexity in Model Ecosystems*. Princeton University Press, Princeton, N.J.

REFERENCES

May, R. M. 1977. *Theoretical Ecology: Principles and Applications*. Saunders, Philadelphia.

May, R. M. 1985. Regulation of population with nonoverlapping generations by microparasites: a purely chaotic system. *Amer. Natur.* 125: 573–584.

May, R. M., and G. F. Oster. 1976. Bifurcation and dynamic complexity in simple ecological models. *Amer. Natur.* 110: 573–599.

Maynard Smith, J. 1964. Kin selection and group selection. *Nature* 201: 1145–1147.

Maynard Smith, J. 1976. Group selection. *Quart. Rev. Biol.* 51: 277–283.

Maynard Smith, J. 1982. *Evolution and the Theory of Games*. Cambridge University Press, Cambridge.

Morse, D. H. 1980. *Behavioral Mechanisms in Ecology*. Harvard University Press, Cambridge, Mass.

Nicholson, A. J. 1954. An outline of the dynamics of animal populations. *Austral. J. Zool.* 2: 9–65.

Parker, G. A. 1983. Arms races in evolution: an ESS to the opponent-independent costs game. *J. Theor. Biol.* 101: 619–648.

Petrusewicz, K. 1957. Investigation of experimentally induced population growth. *Ekol. Pol.* 5: 281–309.

Petrusewicz, K. 1963. Population growth induced by disturbance in the ecological structure of the population. *Ekol. Pol.* 11: 87–125.

Pielou, E. C. 1977. *Mathematical Ecology*. Wiley, New York.

Poole, R. W. 1974. *An Introduction to Quantitative Ecology*. McGraw-Hill, New York.

Pulliam, H. R., and T. Caraco. 1984. Living in groups: is there an optimal group size? In J. R. Krebs and N. B. Davies, eds., *Behavioral Ecology: An Evolutionary Approach*, 122–147. Blackwell, Oxford.

Readshaw, J. L., and W. R. Cuff. 1980. A model of Nicholson's blowfly cycles and its relevance to predation theory. *J. Anim. Ecol.* 49: 1005–1010.

Ritte, U. 1969. Floating and sexuality in laboratory populations of *Hydra littoralis*. Ph.D. thesis, University of Michigan, Ann Arbor.

Roff, D. A. 1975. Population stability and the evolution of dispersal in a heterogeneous environment. *Oecologia* 19: 217–237.

Ross, M. A., and J. L. Harper. 1972. Occupation of biological space during seedling establishment. *J. Ecol.* 60: 77–88.

Roughgarden, J., S. Pacala, and J. Rummel. 1984. Strong present-day competition between the *Anolis* lizard populations of St. Maarten (Neth. Antilles). In B. Shorrocks, eds., *Evolutionary Ecology*, 203–220. Blackwell, Oxford.

Rubenstein, D. I. 1981a. Individual variation and competition in Everglade pygmy sunfish. *J. Anim. Ecol.* 50: 337–350.

Rubenstein, D. I. 1981b. Population density, resource patterning and territoriality in the Everglade pygmy sunfish. *Anim. Behav.* 29: 155–172.

Schoener, T. W. 1973. Population growth regulated by intraspecific competition for energy and time: some simple representation. *Theor. Popul. Biol.* 4: 56–84.

Slobodkin, L. B. 1964. Experimental populations of *Hydrida*. *J. Anim. Ecol.* 33 (Suppl.): 131–148.

Soberon, J. 1986. The relationship between use and suitability of resources and its consequences to insect population size. *Amer. Natur.* 127: 338–357.

Sokoloff, A. 1974. *The Biology of Tribolium*. Vol. 2. Clarendon Press, Oxford.

Southwood, T.R.E. 1962. Migration of terrestrial arthropods in relation to habitat. *Biol. Rev.* 37: 171–214.

Stewart, F. M., and B. R. Levin. 1973. Partitioning of resources and the outcome of interspecific competition: a model and some general considerations. *Amer. Natur.* 107: 171–198.

Tamarin, R. H., L. M. Reich, and C. A. Moyer. 1984. Meadow vole cycles within fences. *Can. J. Zool.* 62: 1796–1804.

Taylor, L. R., and R.A.J. Taylor. 1977. Aggregation, migration and population mechanics. *Nature* 265: 415–421.

Taylor, L. R., and R.A.J. Taylor. 1978. The dynamics of spatial behaviour. In F. J. Ebling and D. M. Stoddart eds., *Population Control by Social Behaviour*, 181–212. Institute of Biology, London.

Taylor, L. R., and R.A.J. Taylor. 1983. Insect migration as a paradigm for survival by movement. In I. R. Swingland and P. J. Greenwood, eds., *The Ecology of Animal Movement*, 181–214. Claredon Press, Oxford.

Terman, C. R. 1980. Behavior and regulation of growth in laboratory populations of prairie deermice. In M. N. Cohen, R. S. Malpass, and H. G. Klein eds., *Biosocial Mechanisms of Population Regulation*, 23–36. Yale University Press, New Haven, Ct.

Thomas, W. R., M. J. Pomerantz, and M. E. Gilpin. 1980. Chaos, asymmetric growth and group selection for dynamic stability. *Ecology* 61: 1312–1320.

Turner, M. D., and D. Rabinowitz. 1983. Factors affecting frequency distribution of plant mass: the absence of dominance and suppression in competing monocultures of *Festuca paradoxa. Ecology* 64: 469–475.

Uchmanski, J. 1983. The effect of emigration on population stability: a generalization of the model of regulation of animal numbers, based on individual differences. *Oikos* 41: 49–56.

Uchmanski, J. 1985. Differentiation and distribution of body weights of plants and animals. *Phil. Trans. London Roy. Soc.* B, 310: 1–75.

Vandermeer, J. 1981. *Elementary Mathematical Ecology*. Wiley, New York.

Van Valen, L. 1971. Group selection and the evolution of dispersal. *Evolution* 25: 591–598.

Van Valen, L. 1973. Pattern and the balance of nature. *Evol. Theor.* 1: 31–49.

Varley, C., G. R. Gradwell, and M. P. Hassell. 1973. *Insect Population Ecology: An Analytical Approach*. University of California Press, Berkeley.

Wallace, B. 1968. Polymorphism, population size and genetic load. In R. C. Lewontin, ed., *Population Biology and Evolution*, 87–105. Syracuse University Press, Syracuse, N.Y.

Wallace, B. 1975. Hard and soft selection revisited. *Evolution* 29: 465–473.

Wallace, B. 1977. Automatic culling and population fitness. *Evol. Biol.* 10: 265–276.

Wallace, B. 1981. *Basic Population Genetics*. Columbia University Press, New York.

Wallace, B. 1982. Phenotypic variation with respect to fitness: the basis for rank-order selection. *Biol. J. Linn. Soc.* 17: 269–274.

Weiner, J., and O. T. Solbrig. 1984. The meaning and measurement of size hierarchies in plant populations. *Oecologia* 61: 334–336.

Werner, E. E., and J. F. Gilliam. 1984. The ontogenic niche and the species interaction in size structured populations. *Ann. Rev. Ecol. Syst.* 15: 393–425.

Werner, P. A. 1975. Prediction of fate from rosette size in teasel (*Dipsacus fullonum* L.). *Oecologia* 20: 197–201.

Wethey, D. S. 1983. Intrapopulation variation in growth of sessile organisms: natural populations of the intertidal barnacle *Balanus balanoides*. *Oikos* 40: 14–23.

Whitham, T. G. 1978. Habitat selection by *Pemphigus* aphids in response to resource limitation and competition. *Ecology* 59: 1164–1176.

Whitham, T. G. 1980. The theory of habitat selection examined and extended using *Pemphigus* aphids. *Amer. Natur.* 115: 449–466

Wilbur, H. M. 1976. Density-dependent aspects of metamorphosis in *Ambystoma* and *Rana sylvatica*. *Ecology* 57: 1289–1296.

Wilbur, H. M. 1980. Complex life cycles. *Ann. Rev. Ecol. Syst.* 11: 67–93.

Wilbur, H. M., and J. P. Collins. 1973. Ecological aspects of amphibian metamorphosis. *Science* 182: 1305–1314.

REFERENCES

Wilson, D. S. 1980. *Natural Selection of Populations and Communities*. Benjamin/Cummings, Menlo Park, Calif.

Wynne-Edwards, V. C. 1962. *Animal Dispersion in Relation to Social Behaviour*. Oliver and Boyd, Edinburgh.

Wyszomirski, T. 1983. A simulation model of growth of competing individuals of a plant population. *Ekol. Pol.* 31: 73–92.

Ziołko, M., and J. Kozłowski. 1983. Evolution of body size: an optimization model. *Mathem. Biosci.* 64: 127–143.

Author Index

Note: The Author Index includes those cited in figure captions as well as in the text. Also included are the names of third or fourth authors cited simply as et al. in the text.

Aitchison, J., 49
Albon, S. D., 90
Allee, W. C., 2, 132

Bach, C. E., 203
Bacon, P. J., 88, 96
Begon, M., 106
Bell, G., 195
Bellows, T. S., Jr., 35, 39, 42
Brace, R. C., 111
Braun-Blanquet, J., 2
Brown, J.A.C., 49
Bryant, D. M., 58

Cadwell, L. D., 187
Caraco, T., 166, 167, 172
Charnov, E. L., 148
Ciesielska, M., 195
Clutton-Brock, T. H., 90
Cohen, M. N., 135
Colinvaux, P. A., 3
Collins, J. P., 47, 50
Comins, H. N., 156, 157, 158, 160
Cooper, W. S., 66, 68, 69, 70, 72
Cuff, W. R., 38

Darwin, C., 1
Davies, N. B., 204
Dawkins, R., 3, 72
Dawson, P. S., 110
de Jong, G., 19, 51, 197, 198
Dewsbury, D. A., 119
Donald, C. M., 41

Emerson, A. E., 2, 132
Errington, P. L., 204

Falconer, D. S., 82, 84
Finerty, J. P., 148
Fleischer, R. C., 84
Fretwell, S. D., 170, 171
Fujii, K., 61

Gentry, J. B., 187
Gill, D. E., 151
Gilliam, J. F., 92
Gilpin, M. E., 92, 125, 151, 155
Gliwicz, J., 96
Gould, S. J., 1
Gradwell, G. R., 43
Grant, P. R., 155
Grinnell, J., 189, 204
Guinness, F. E., 90
Gurney, W.S.C., 38, 64
Gustafsson, L., 82

Hairston, N. G., 125
Hamilton, W. D., 147, 156, 157, 158, 160
Harper, D.G.C., 171
Harper, J. L., 35, 36, 37, 41, 42, 47, 58, 195
Hassell, M. P., 19, 35, 107, 197, 198
Hewlett, P. S., 8, 43, 48
Heywood, J. S., 85
Holling, C. S., 55, 63
Horn, H. S., 190
Hubbell, S. P., 89
Hughes, T. P., 88, 96

Ingram, C., 58

Jagers, P., 10
Johnson, R. F., 84

Kaplan, R. H., 66, 68, 70, 72
Keller, B. L., 187
Kepteyn, J. C., 49
Kimmel, M., 56
Kira, T., 41, 47, 49, 50, 57
Klein, H. G., 135
Klitz, W. J., 84
Knapton, R. W., 119
Kositzin, V. A., 2
Kot, M., 94
Koyama, H., 47, 49, 50, 57
Kozłowski, J., 17, 58
Krawczyk, J., 113, 114
Krebs, C. J., 187, 188
Krebs, J. R., 119, 154, 204

Laskowski, R., 94, 95
Lawton, J. H., 38
Le Cren, I. D., 42
Lerner, I. M., 84
Leslie, P. H., 96
Levin, B. R., 15
Levin, D. A., 85
Levins, R., 17, 66, 150, 155
Lewontin, R. C., 65, 81, 86
Lidicker, W. Z., Jr., 153, 154, 189
Lloyd, J. A., 140
Łomnicki, A., 18, 33, 54, 55, 63, 64,
    113, 114, 125, 129, 179, 180, 182
Lotka, A. J., 2, 10, 17
Lucas, H. L., Jr., 170

Malpass, R. S., 135
Malthus, T., 1
May, R. M., 12, 19, 30, 33, 147,
    156, 157, 158, 160, 197, 198, 202
Maynard Smith, J., 65, 66, 121,
    125, 129, 151, 155, 159
Morse, D. H., 119
Moyer, C. A., 188

Nicholson, A. J., 37, 38, 42, 43, 44,
    110
Nisbet, R. M., 38, 64

Ogawa, H., 41, 47
Ombach, J., 63
Oster, G. F., 30

Pacala, S., 187
Park, O., 2, 132
Park, T., 2, 132
Parker, G. A., 121, 122, 123, 127
Pavey, J., 111
Perrins, C., 154
Petrusewicz, K., 134, 141, 142
Pielou, E. C., 13
Plackett, R. L., 8, 48
Pomerantz, M. J., 99
Poole, R. W., 13, 87
Pulliam, H. R., 166, 167, 172

Quetelet, A., 1

Rabinowitz, D., 50
Readshaw, J. L., 38
Reich, L. M., 188
Ritte, U., 179, 180
Roff, D. A., 156
Ross, M. A., 58
Roughgarden, J., 187
Rubenstein, D. I., 13, 14, 110, 120,
    190
Rummel, J., 187

Sakazaki, N., 41, 47
Schaffer, W. M., 94
Schmidt, K. P., 2, 132
Schoener, T. W., 15
Slobodkin, L. B., 125, 178, 179, 180
Smith, F. E., 125
Soberon, J., 125
Sokoloff, A., 115, 185
Solbrig, O. T., 52
Southwood, T.R.E., 151
Stewart, D., 1
Stewart, F. M., 15

Tamarin, R. H., 187, 188
Taylor, L. R., 189
Taylor, R.A.J., 189
Terman, C. R., 134, 140

AUTHOR INDEX

Thomas, W. R., 99
Turner, M. D., 50

Uchmanski, J., 46, 47, 57, 58, 63, 102

Vandermeer, J., 30
Van Valen, L., 125, 155, 156, 160
Varley, C., 43.
Volterra, V., 2, 17

Wallace, B., 66, 74, 75, 77
Weiner, J., 52

Werner, E. E., 92
Werner, P. A., 88, 89
Wethey, D. S., 57
Whitham, T. G., 116, 117, 173
Wilbur, H. M., 39, 40, 42, 47, 50, 91
Wilson, D. S., 32, 125, 127, 130
Wynne-Edwards, V. C., 2, 40, 125, 153, 189, 204
Wyszomirski, T., 59, 60

Ziołko, M., 58

Subject Index

age, stable distribution of, 10, 11
age-dependent reproduction and
    survival, 73, 86–91, 96
asymmetric competition, 106, 110
asymmetric game, 149

barnacles (body size), 56
binomial distribution, 51
blowfly: larvae (survival and
    scramble competition), 37, 38, 43,
    44; populations, 38
body weight, distribution of:
    general, 46; competition for space,
    59, 60; deterministic body growth,
    57–59; empirical data, 46–48, 50,
    56, 58; predictor of individual
    fate, 55, 57, 88–90, 96; stochastic
    body growth, 53–57; theoretical
    explanations, 48–53, 56, 57. *See
    also* size-dependent reproduction
    and survival

cannibalism, 110, 185
carrying capacity, 11, 18, 125
cell cycle, 10
clone, 3, 12, 128, 129, 148, 182
clover (constant final yield), 41
clumped dispersion, 19, 60, 198
coin-flipping strategy, 70–72
collared flycatcher (fitness
    heritability), 82
colonization: ability, 17; rate, 16, 17
competing group, 91, 92, 95
competition: asymmetric, 106, 110;
    exploitative, 110, 118; for light,
    107, 108, 110, 111; for space, 59;
    interference, 110; skewed
    distribution of body weights, 56.
    See also contest competition,
    scramble competition
complex life cycles, 91

conditioning, effects on: floating of
    Hydra polyps, 180, 181;
    population growth, 133
confined populations: field, 186–
    188; laboratory, 134, 178–180,
    183–185
constant final yield, law of, 41
contest competition: arms-races,
    120–123; asymmetric competition,
    106, 110; cannibalism, 110;
    definition of, 45, 107–111; egg-
    laying by flour beetles, 113–115;
    emigration, 166–169; forest trees,
    107, 108; hierarchies, 118, 119;
    original definition, 43, 44;
    population effects, 111–115;
    reproductive success among
    aphids, 116–118; resource
    partitioning, 45, 63, 107–109; soft
    mortality, 109; territoriality, 113,
    118–120
continuous models, 100–105
cucumber beetle (population
    density), 203

decision theory, 66–69
despotic distribution, 170–173
determinate duration of local
    habitats, 152, 153, 183
deterministic body growth, 57–59
developmental homeostasis, 84
discrete models, 100–105

ecosystem stability, 202
emigration: asymmetric game, 149;
    contest competition, 166–169,
    172, 173; egg-laying by flour
    beetles, 113–115; from groups of
    related individuals, 145–149; from
    groups of unrelated individuals,
    143–145; models of, 153;

population regulation by, 153, 154, 161, 166, 189, 201, 203; without unequal resource partitioning, 155–161; with unequal resource partitioning, 161–166
Enigma of Balance, 125, 126
Everglades pygmy sunfish: contest competition, 120; individual differences, 14; territorial behavior, 110, 120
evolutionarily stable fraction of migrants, 148, 157, 160, 164
evolutionarily stable strategy (ESS): arms-races, 120–123; asymmetric, 149; emigration 143–150; existential game, 126, 128; reproduction in heterogeneous environments, 192, 193; self-regulation, 126–132
evolution of emigration, 143–150, 155–169
extinction. See population extinction, population persistence
extinction probability, 191

fish (territoriality), 42, 110, 120
flour beetles: egg-laying, 113–115, 118; free and confined populations, 183–186; interference competition, 110, 112; larva survival, 94, 95
food used, 181, 185
free distribution, 170–172
free populations, 180–186
fruit flies (population stability), 99, 100
functional response of predators, 55, 63

gall aphids: contest competition, 116–118; shoving contest, 118, 149
genetic variation: general, 65; individual success, 81–84; population stability, 85
geometric series, 64

Gini coefficient, 52
group selection, 2, 3, 100, 125, 127, 151, 155, 156, 160, 189

hard selection, 75. See also soft selection
heritability of fitness characters, 82–84
heterogeneity. See spatial heterogeneity
hierarchy. See social hierarchy
holistic approach, 4, 134, 135
homozygosity: phenotypic variability, 84, 85; developmental homeostasis, 84
Hydra polyps (emigration), 178–182

ideal free distribution, 170
immigration, 151, 154, 155
indeterminate duration of local habitats, 152, 153, 181, 183
individual, uniqueness of, 3
individual properties, 6, 13
individual resource intake, 21. See also resource partitioning among individuals
individual susceptibility, 9
individual variation: adaptation, 66–72; stability, 32–34

kin selection, 128–131, 147, 148

life history strategies, 136
life stages, 91–95, 112
lifetime reproductive success, 90, 91, 142
local habitat, 150–153
local population, 150, 151
logarithmic normal distribution, 8, 9, 21, 49, 50
logistic equation, 4, 11–15, 17
Lorenz curve, 52

macroheterogeneity, 194, 197. See also spatial heterogeneity
maintenance cost, 23, 25, 112
maize (population dynamics), 35

mallards (free distribution), 171, 172
maximum individual resource
    intake. *See* satiation level
metapopulation, 150, 151
microheterogeneity, 194–197. *See*
    also spatial heterogeneity
microorganisms (population
    dynamics), 10
migration. *See* emigration
monopolization of resources, 34,
    110, 118
mortality, differential, 73–77, 80

normal distribution, 51

optimal decision theory, 66
overlapping generations, 95–100
overproduction of offspring, 190

persistence. *See* population
    persistence
plants (cultivated versus wild), 36,
    37, 41
population decline, 8, 11
population dynamics: discrete versus
    continuous models, 100–105; four
    versions of the model, 25–34;
    general model, 23, 25;
    overlapping generations, 95–100
population extinction, 35, 38, 40,
    99, 100, 136, 156, 159, 160
population field studies, limitation
    of, 174–176
population growth: artificially
    induced, 141, 142; effect of
    conditioning on, 133; limited, 11–
    15; limited in discrete places, 15–
    19, 190; unlimited, 5–11;
    unlimited, of microorganisms, 10
population laboratory studies,
    limitations of, 176–178
population persistence: general, 30–
    32; for four versions of the model,
    32–34; for overlapping
    generations, 98–100; resource
    depletion, 40

population self-regulation:
    definition, 124; game theory, 126–
    132; group selection, 125; in
    laboratory, 132–136; unequal
    resource partitioning, 136–142
population stability: general, 12, 18,
    19, 30–32; for four versions of the
    model, 32–34; for overlapping
    generations, 97–100
prairie grass (size distribution), 50
presaturate emigration, 154, 155,
    161, 166
proportionate effect, theory of, 49,
    54

quantal response, analysis of, 8, 48

r and K selection, 190
random dispersion, 60
rank, individual, 21, 23, 62, 64, 80,
    84, 137, 138
reductionist approach, 4, 134, 135
refraining from reproduction, 139,
    141, 142
reproductive rate, net, as the mean
    of random variable, 7
reproductive success of emigrants,
    164
resource depletion, 40, 136
resource inflow, 21
resource partitioning among
    individuals: general, 20–25;
    formal properties of, 62–64; four
    versions, 25–30; in relation to
    body weight distributions, 61–63;
    laboratory and field data, 34–42

satiation, 53, 54, 56
satiation level, 23, 53
scramble competition, 37, 43, 44,
    166, 167. *See also* contest
    competition
selection: artificial, 83; hard, 75; r
    and K, 190. *See also* soft selection
self-regulation. *See* population self-
    regulation
self-thinning, 37

size-dependent reproduction and
survival, 88–90, 96
skewed distribution of: body weight,
47, 49, 50, 52, 56, 57, 60;
individual susceptibility, 9
social hierarchy, 21, 118, 119, 137,
138, 141, 142
soft selection: general, 74, 76–81;
against recessive homozygotes,
77–79; against self-regulation,
125; age-dependent mortality, 87;
contest competition, 109;
differential mortality, 73–77
song sparrow (territoriality), 119
spatial heterogeneity: general, 194;
direct relation to stability, 197–
199; ecosystem stability, 201–203;
genetic variability, 81, 195;
macroheterogeneity, 194, 197;
measure of, 199, 200;
microheterogeneity, 194–197;
species diversity, 199–201;
variability of individual success,
195–197

species diversity, 199–201, 202
stability. *See* population stability,
ecosystem stability
stochastic body growth, 53–57
superorganism, 2–4
survival probability, as an
individual property, 6, 13

tadpoles (survival), 39
teasel, common (age and size
dependence), 88–90
territoriality, 42, 113, 118–120
threshold rank, 163
trait-group, 127, 130
trout fry (survival), 42

variation as an adaptation, 66–72
variation coefficient, 51
voles (confined populations), 187,
188

$y(x)$ function, 21–23, 56, 62–64. *See
also* resource partitioning among
individuals

LIBRARY OF CONGRESS CATALOGING-IN-PUBLICATION DATA

Łomnicki, Adam, 1935–
Population ecology of individuals.

(Monographs in population biology; 25)
Bibliography: p.   Includes indexes.
1. Population biology.   2. Ecology. I. Title. II. Series.
QH352.L66   1988   591.52'48   87–3439
ISBN 0–691–08471–8 (alk. paper)
ISBN 0–691–08462–9 (pbk.)